Information Technology Standards and Standardization: A Global Perspective

Kai Jakobs
Technical University of Aachen, Germany

IDEA GROUP PUBLISHING
Hershey USA • London UK

Senior Editor: Mehdi Khosrowpour
Managing Editor: Jan Travers
Copy Editor: John Syphrit
Typesetter: Tamara Gillis
Cover Design: Connie Peltz
Printed at: BookCrafters

Published in the United States of America by
        Idea Group Publishing
        1331 E. Chocolate Avenue
        Hershey PA 17033-1117
        Tel: 717-533-8845
        Fax: 717-533-8661
        E-mail: jtravers@idea-group.com
        Website: http://www.idea-group.com

and in the United Kingdom by
        Idea Group Publishing
        3 Henrietta Street
        Covent Garden
        London WC2E 8LU
        Tel: 171-240 0856
        Fax: 171-379 0609
        Web site: http://www.eurospan.co.uk

Library of Congress Cataloging-in-Publication Data

Jakobs, Kai, 1957-
        Information technology standards and standardization : a global perspective /
        Kai Jakobs.
            p. cm.
        Includes bibiliographical references and index.
        ISBN 1-878289-70-5 (paper)
            1. Information technology--Standards. I. Title.

        T58.5 .J35 1999
        004'.0218--dc21   99-045046

British Cataloguing in Publication Data
A Cataloguing in Publication record for this book is available from the British Library.

Information Technology Standards and Standardization: A Global Perspective

Table of Contents

Preface

"Standards are not only technical questions. They determine the technology that will implement the Information Society, and consequently the way in which industry, users, consumers and administrations will benefit from it."

You can hardly put it more to the point than this quote taken from a document published by the Commission of the European Union on 'Standardization and the Global Information Society'. Despite all criticism, and despite descriptions such as 'hampering progress' or 'trailing behind the market,' standards remain the sine-qua-non in virtually all fields of technology, and especially in information technology (IT).

Our world is becoming networked. The envisaged Global Information Infrastructure (GII), for instance, is going to have a profound impact as the major enabler of the frequently predicted move from an industrial society to the information society. In the meantime, initiatives towards national or regional information infrastructures are gaining momentum. This holds particularly for the US, the Pacific Rim and Europe. Likewise, major developments may be observed in the domestic sector, where stand-alone computers are bound to vanish sooner or later. ISDN interconnectivity, and particularly access to the Internet are increasingly commonplace. It may take some time, but ultimately almost every organisation, company, school and household will be interconnected. Or so they say; this frequently evoked development can only happen if globally agreed standards will be available, upon which this infrastructure can be based.

Perhaps the most fascinating quality to be associated with standards and standardization, and certainly their most problematic characteristic is the multitude of dimensions that need to be considered. These include, but are definitely not limited to the economics of standards, standardization policies, intellectual property rights, the overall structure of standards setting processes and whether their output manages to address today's needs. Obviously, the technical quality of the specifications, and whether or not they find a window of opportunity play a role as well. Likewise, corporate strategies arc vital to a standard's prospects in the market. Maybe less obvious, the setting whithin which a standards based system is to be employed also plays a decisive role; that is, these systems need to blend in seemlessly with their respective environment to be acceptable and successful.

Stakeholders in the standards setting process are at least as diverse, ranging from governments to, ultimately, the individual end users. Large multi-national vendors/manufacturers have specific requirements on standards and the standards setting process, which may well contradict those of small or medium sized firms, which in turn are typically very different from the needs and expectations of user companies. The human end-user is again a totally different issue altogether.

Reading the chapters you will find that the approaches to the topic differ considerably, as do the views of the individual authors, their respective explicit or implicit definition of the term 'standard' and, indeed, their styles of writing (I certainly didn't want to impose a uniform overall structure or style). Little wonder, though, given the immense diversity of facets and problems to be associated with standards and standardization, and their frequently controversial character. The contributions you will find range from the introductory to the very specific; they cover different aspects of standards setting processes as well as corporate issues and economics, and present selected case studies.

You will also note that some topics are discussed in more than one chapter; these include particularly the issue of 'speed' in standards setting and the question of how to address Intellectual Property Rights (IPRs). This is not coincidental—at least (but not only) in my view these problems are the most critical ones with respect to the establishment of a standardization infrastructure. This infrastructure will have to be capable of coping with the needs and requirements of a technology (and its users) whose ever increasing importance, together with its sheer pace of development and its ever shorter life cycles, necessitates a rethinking of today's standards setting process, and may well require a process very different from the one we have today.

Acknowledgments

This was my first attempt at book editing, and certainly it was more work than I had anticipated. Yet, I don't want to think of how bad things would have been without the support from many different sides. First of all, I have to praise the reviewers, who did a tremendous job (some of them less than exactly voluntarily). Luckily, my boss, Prof. Spaniol, accepted my view that editing a book is part of my job, so I could do the bulk of the work during —somewhat extended—office hours; my thanks go to him as well. The good folks at IGP never really tried to push me, and patiently supported their new novice editor. Thanks.

And finally, of course, thanks to Martina, who certainly didn't like my getting out of bed in the middle of the night to go to work.

Chapter I

Knowledge Age Standards: A brief introduction to their dimensions

Yesha Y. Sivan
Tel Aviv University and
The Knowledge Infrastructure Laboratory, Ltd.

OVERVIEW

The diverse uses of "standards" define the goal of this work, namely, to develop a general framework of standards and to reflect on the process and outcome of the development of the framework. My intention is to devise a theoretical framework that may be translated into practice at some future point.

The principle outcome is a framework of standards that includes five dimensions: Level, Purpose, Effect, Sponsor, and Stage, each of which contains five categories that together define the dimension. The dimensions show:

- how standards can be produced and used by entities from different Levels (individual, organizational, associational, national, and multinational);
- how they can have one or more Purposes (simplification, communication, harmonization, protection, and valuation);
- how they can cause diverse Effects (constructive, positive, unknown, negative, and destructive);
- how they can be developed by different Sponsors (devoid, nonsponsored, unisponsored, multisponsored, and mandated); and
- how they can be in different Stages (missing, emerging, existing, declining, and dying).

In presenting the framework, the chapter also touches on the roles of standards in the industrial age, their potential roles in the knowledge age, and current turmoil in the standards community. It includes reflections on designing and judging the framework.

BACKGROUND: SPHERE OF STANDARDS IN THE INDUSTRIAL AGE

Every aspect of our life is supported and often controlled by standards. Consider, for example, the book you are now reading. It has a table of contents (a common standard for quick access); it has page numbers (another quick access device); it uses a standard language, a standard font, and a standard paper size. In the making of this work, both directly or indirectly, I have used dozens of other standards, among them: the Postscript page description language, the Internet, the Harvard on-line library system, the QWERTY keyboard, the Microsoft Word program, and many more.

Assume that you are sitting in a typical kitchen of a typical home anywhere in the industrialized world. Look around you. All electrical appliances share the same electric current. Need a fan? Move it from another room, plug it in, and enjoy the cool breeze. Want some music? No problem; grab any tape and a tape player, put the tape in the player, press "play," and enjoy the sounds. Notice that you can take any tape from any vendor, record on it in your own tape player, and replay it in any other tape player, anywhere in the world.

Assume that you are in a car. Look around you. First and foremost is the issue of fuel, which you can get in any fuel station on any street or in any city. Look at your tires. You can choose any kind, as long as they match the standard specifications of your car. Consider the plate number, registration, and mandated insurance, or the traffic signs, directional lights, emission standards, and the radio. All involve some standards.

What were the roles standards played in the industrial age? As the above examples suggest, standards played diverse roles. One researcher (Gaillard, 1934; cited in Verman, 1973, p. 22) offered the following laundry list:

A standard is a formulation established verbally, in writing or by any other graphical method, or by means of a model, sample, or other physical means of representation, to serve during a certain period of time for defining, designing or specifying certain features of a unit or basis of measurement, a physical object, an action, a process, a method, a practice, a capacity, a function, a duty, a right, a responsibility, a behavior, an attitude, a concept or a conception, or a combination of any of these, with the object of promoting economy and efficiency in production, disposal, regulation and/or utilization of goods and services, by providing a common ground of understanding among producers, dealers, consumers, users, technologists and other groups concerned.

While comprehensive, this list has no zest, charm, or appeal. Such a definition, although considered "a classic [that] has served the profession for many decades" (Verman, p. 22), often deters people because it does not engage them in any meaningful way. What I needed personally, and what I feel others need in order to embrace the concept of standards, is a strong evocative image that will capture the critical facets of the phenomenon of standards.

In searching for this evocative metaphor for standards in the industrial age, I have created the "cultures and sphere" image. In this image, the industrial age is represented as two cultures that operate in a sphere. The first culture, which I call the culture of "technology," has to do with the invention of tools that allow men and women to produce more and ultimately consume more. The second culture, which I call the culture of "business," has to do with the management of the production, marketing, and finances that move technologies from the labs into the markets. The sphere represents the many standards that have oiled the industrial age and facilitated the smooth interaction between business and technology.

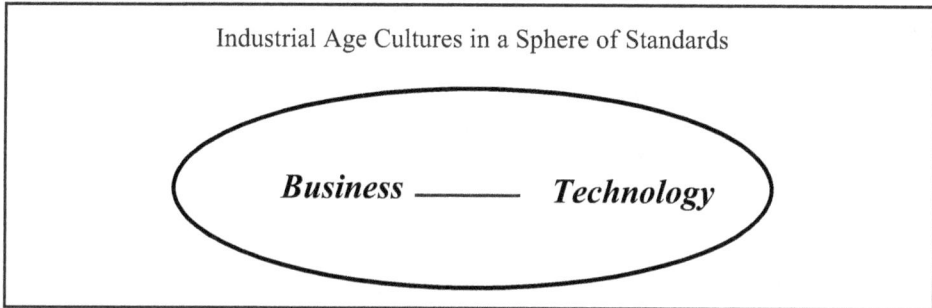

Industrial Age Cultures in a Sphere of Standards

Business ———— Technology

Figure 1 - Industrial Age Sphere of Standards

This model, like any other model, captures what I consider the most important facets of the phenomenon. In this case, the cultures and sphere image was designed to capture in a succinct, evocative, and memorable way the roles standards played in the industrial age. As to terminology, I use the word "culture" to mean a mindset, a particular way to look at the world. For example, a technology person is interested in artifacts, how they work, how they can work better, and the like; a business person, on the other hand, is interested in costs, processes, control structures, and the like. I use the word "sphere" to mean an environment, in our case mostly hidden, that facilitates the smooth operation of the cultures.

From my limited use of this image, I can say that it catches people's attention, and it seems to prompt ample questions about the roles of standards and their specific meaning in relation to business and technology. Since this image describes the industrial age, the next logical question has to do with the roles standards are expected to play in the post-industrial age, an age that is often called the "knowledge age."

BIGGER SPHERE OF STANDARDS IN THE KNOWLEDGE AGE

The dawn of the knowledge age is here, and we — as individuals, organizations, and nations — already feel its emerging challenges. Armed only with industrial-age frameworks, we have to deal with daily television scenes that blend newsmakers with news reporters; round-the-clock, round-the-world, computer-controlled financial activities; and industries struggling to carve their own future in an ever-changing world.

Various pundits (Toffler, 1971, 1974, 1981, 1990; McCluhan, 1964; Naisbitt, 1982, 1990) claim that the future holds an intense interaction with knowledge. To deal with this knowledge, we use personal hand-held information managers, interactive cable television, cellular telephones, and individual newspapers. They are all parts of society's response to the glut of available knowledge, a glut that is marked by an image-intensive, fast-paced culture symbolized by international global names like Big Bird (from Sesame Street), Mario (from Nintendo), and Butthead (from MTV).

The human race enters the knowledge age still equipped with the same innate processing power that served the prehistoric person. More and more, we find ourselves overwhelmed by the complexities of modern life. Here are a few examples:

Personal financial management has long gone beyond the reach of the lay person. Stocks, bonds, futures, options, and other money-making (and more so, money-losing) terms have led most of us to seek professional help in managing our savings. Reading a bank statement is like looking at an encrypted message — you know some of the letters but you really don't understand the message. If we do find a mistake and attempt to have it corrected, the tellers too often respond with evasive maneuvers like, "the computer is down" or "well, call our adjustment department."

The growth of knowledge due to information technologies has caused many of us to confront "information gridlock." The ability to generate reports, papers, and data using new technologies has often caused information to get lost, misplaced, renamed, or erased. Electronic communication systems, if not managed carefully, may overwhelm our already full lives (an occasional concern to some electronic mail users who watch their mailboxes jam with dozens of electronic mail messages).

Another challenge, health care, is beginning to dominate both the first and third worlds. And while research is focused on new medical treatments, there are growing discussions about prevention in the form of knowledge distribution. Fueled by AIDS, "education" is now advocated as a key factor to prevent many medical (and social) nightmares. For example, compare $135, the cost calculated a few years back to reduce the chances of teenage pregnancy through education of teenagers about sex and pregnancy, against the much larger cost of 20 years of public assistance to a child born to a teenage parent (Brady, Taylor, and Willwerth, 1990).

These are just a few of the new challenges that the knowledge age has brought upon us. For some people, especially those in school who will become adults in the knowledge age, these challenges have strong ramifications. To find a job, one will have to monitor all the information channels; to invest or borrow, one will have to educate oneself about a complex set of rules; and to make health-care decisions, one will have to master statistics, decision-making, and the sociology of doctors' prestige.

Now we can turn back to our pending question about the roles of standards in the knowledge age. It seems, at least from the above description of the knowledge age, that the cultures of business and technology will be joined by a third culture, which can be called a culture of "knowledge." And what will be the roles of standards? My prediction is that standards, which facilitated the interactions between the cultures of the industrial age, will also facilitate the interaction among

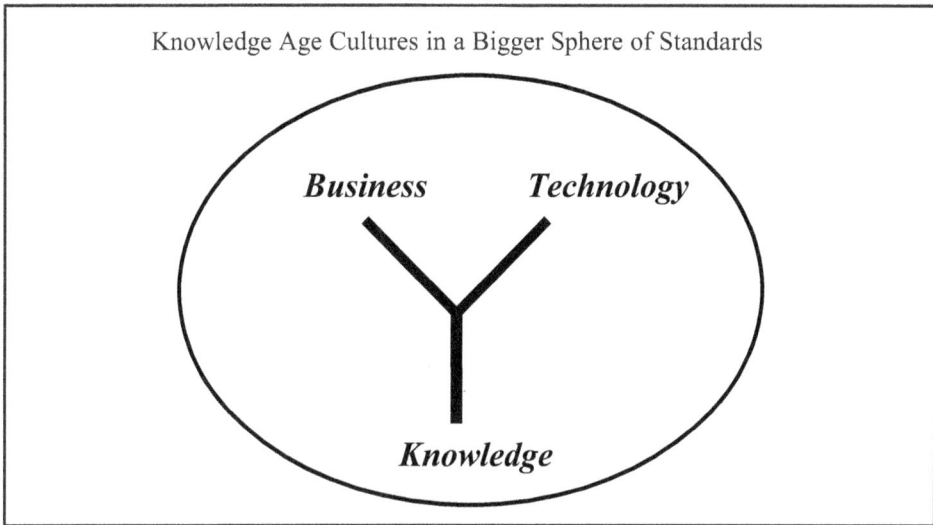

Figure 2 - Knowledge Age Sphere of Standards

the three cultures of the knowledge age.

What do I mean by "knowledge"? Simply put, knowledge is the direct result of learning by individuals, organizations, and even nations. Here are some familiar examples of human learning: As infants, we learn to talk, walk, and listen, as well as to cry for attention. At school, we learn to read, write, and calculate, as well as to cheat, yell, and gossip. Later, in higher education, we learn how to write papers, conduct experiments, and deal with others, as well as how to talk about something without fully understanding it. As working adults, we learn to become practitioners of our callings and to present ourselves to others, as well as to work overtime deliberately. And as we grow older, we unlearn and relearn how to plan and consider, deal with failure, redefine goals ex post facto, and to deal with sickness, as well as how to enjoy life in places like Florida.

Why a "culture of knowledge"? Because knowledge has always given us the know-how, skills, attitudes, and dispositions to deal with the complexities of the world. Because in the knowledge age, the transfer of knowledge in formal schools, universities, boot camps, workshops, seminars, one-hour presentations, self paced learning environments, and other areas will play a growing role in the development of individuals, organizations, and nations. Because learning, which is the transfer of knowledge, is hailed by pundits like Michael Dertouzos (1989) with his "Neglect of Human Resources"; Peter Senge (1990), with his "Learning Disciplines"; and Peter Drucker (1980), with his "Knowledge Worker," as *the* tool for dealing with the complexities of the knowledge age. As *Time* magazine (Lemonick, 1992) succinctly phrased it, "You either learn or perish."

In summary, we know that standards played a major albeit behind-the-scenes role in the industrial age. It is reasonable to assume that in the knowledge age standards will play an even greater role. The cultures of business, technology, and knowledge will demand more standards. As noted by Alvin Toffler (1990), "The fight to control standards . . . . is part of the larger continuing war for the control, routing, and regulation of information. It is a key front in the struggle for power based on knowledge. . . . On every front — scientific, political, economic, and technological — the battle over standards can be expected to intensify as the new system . . . replaces the fast-fading smokestack world of the past" (pp. 139-140).

TURMOIL IN THE STANDARDIZATION COMMUNITY

To grasp fully the scale of the change in the roles of standards, one can look into the current turmoil in the traditional standardization community. This community, which includes the private, national, and international bodies that produced the standards in the industrial age, has to adapt itself to the new roles of standards in the knowledge age.

For example, the International Standards Organization (ISO), in its report *A Vision for the Future: Standards Needs for Emerging Technologies* (1990), claims that traditional industrial-age innovation followed the linear sequence from scientific discovery to applied research and development, followed by production and marketing. This linear sequence, according to the ISO, now needs to evolve into a set of concurrent interactive processes. As a result, the report calls for structural changes in the setting of international standards. This means that while in the industrial age one first created a product and then standardized it, in the knowledge age one often needs the standards before the products. Also in many cases, and especially in information technology industries, compatibility with previous standards is a necessary condition even to enter the market.

In another example, the U.S. Congressional Office of Technology Assessment (OTA), in its report *Global Standards: Building Blocks for the Future* (1992), claims that the "development of a highly competitive global economy, which the United States no longer dominates" (p. 3), will call for more and different global standardization. The report also discusses other aspects of standards in the knowledge age, such as the growth of international standardization efforts and the effect of multinational organizations.

Suddenly the standardization community is called to develop standards in months, rather than in the years it used to take. The official standardization bodies also have to compete with new ad hoc private standardization organizations. In the United States, for example, the "economic competition [between groups that produce standards] is compounded by personality conflicts in the standards setting community" (OTA, 1992, p. 13). One observer, probably a frustrated standards user, complained that "the situation is sheer madness. It has truly gotten out of hand and no longer serves our needs" (p. 13).

From my discussions with members of three major U.S. standardization players, I can testify that this turmoil is quite apparent. At ASTM, the American Society for Testing and Materials, I observed an oiled machine that makes and sells standards, but is extremely fearful about its future. At ANSI, the American National Standards Institute, I observed a well-positioned vessel seeking an experienced captain to overcome years of visionless travel. And at NIST, the

National Institute for Standards and Technology, I observed a well-funded and respected federal agency that is struggling to define its role in relation to the private sector and industrial policy.

In summary, the current turmoil within the standardization community can be seen as a harbinger of the roles of standards in the knowledge age. From these present challenges and tensions we can conclude - with a fair amount of confidence —that standards will play an even bigger role in the knowledge age, as they "transmit information from those who have the knowledge to those who need and can use the knowledge" (Batik, 1992, p. 2).

ORIGIN AND NATURE OF THE FRAMEWORK

Before I could actually start developing the framework, I had to devise a good format for it. Luckily, early in my journey, I found what seemed to be a useful candidate. This format was published in Lal Verman's 1973 seminal work *Standardization: A New Discipline*. In his book, Verman, who was the Director General of the Indian Standards Institute from 1947 to 1955, proposed a three-dimensional standardization space, based on the notion of space "as a logical means of presenting standardization" (Sanders, 1972, p. 16).

Verman's approach to mapping the concept of standards can be best demonstrated by using a simplified example. Suppose we want to understand the concept of "shirts." According to Verman, we first have to find the three major dimensions, or attributes, of shirts. For the sake of the example, let's say that these are the dimensions of color (categories include: black, white, red, yellow, and blue), kind (categories include: fun shirt, work shirt, evening shirt), and size (categories include: small, medium, and large). Then, following Verman, we arrange these dimensions in a three-dimensional space. Each point in the space represents a potential question that one can ask about shirts. For example, who uses a black, long-sleeved, fun shirt? Or, what can we say about work shirts in terms of color or kind? (Note that the dimensions generate questions and not answers.)

Verman himself explained that the three-dimensional space should not be taken in its strict mathematical sense, but more as a way to look systematically at the phenomenon of standards. He also suggested adding more dimensions, which go beyond the spatial representation of the three dimensions. To continue our shirts example, we can add, as a fourth dimension, the shape of the shirt (categories include: long sleeves, short sleeves, has buttons, has pockets).

In general, frameworks like the one proposed by Verman, which attempts to classify a concept systematically, are often used to create a shared map for a concept. Like other maps, they model a complex concept by capturing some of its important dimensions. Their main purpose as models is to "serve as instruments of understanding" (Perkins, 1986, p. 126), which they achieve by highlighting the critical dimensions of the land. As with other frameworks, models, and maps that assist in describing and analyzing their respective domains, a framework of standards should create a common vocabulary, and thus assist in describing and analyzing the domain of standards.

Verman's dimensional approach seemed like a good model. To confirm that, early in my research I tested his approach in the context of a case study. First, based on several sources, I developed a tentative framework of standards. The tentative framework had four dimensions: Domain, Level, Purpose, and Ramification. Each dimension included five sub-dimensions, or categories, which taken together defined the dimension. The dimensions were designed, at the price of some over-simplification and perhaps all-inclusiveness, to identify and analyze standards. Then, to test the applicability of the tentative framework, I used it to examine the nature and roles of standards in one organization (Sivan, 1993a).

This preliminary research confirmed the basic utility of the dimensional approach, but it also suggested that Verman's spatial approach had some limita-tions when it involved more than three dimensions. Further, it seems that a more verbal approach might be more appropriate for a descriptive framework of stan-dards.

The preliminary research also raised again the inherent pitfalls of such a framework. Like all maps (Kent, 1978), a dimensional framework has limitations. Not only can it highlight only parts of the terrain; it may also distort some of the terrain's features. Like the blue line on a map that marks a river that may be dry, certain dimensions that the framework describes in a particular way may look quite different in the real world. In the same way that it is not possible to capture the true color of every river, it is impossible to capture the actual meaning of each dimension in the real world. After all, a map is just a map, and it is not the actual land.

OVERVIEW OF THE FIVE DIMENSIONS

The principal result of this work is a framework for standards that has five dimensions. Each dimension has five categories, which together explicate the dimension.

Dimension 1: Level	Dimension 2: Purpose	Dimension 3: Effect	Dimension 4: Sponsor	Dimension 5: Stage
Individual	Simplification	Constructive	Devoid	Missing
Organizational	Communication	Positive	Nonsponsored	Emerging
Associational	Harmonization	Unknown	Unsponsored	Existing
National	Protection	Negative	Multisponsored	Declining
Multinational	Valuation	Destructive	Mandated	Dying

Table 1 - Summary of the Five Dimensions

The framework can be best illustrated by showing how the five dimensions work in a real context. So, for the purpose of this overview, I would like to give you a taste of the framework. I'm well aware that at this point some of the categories probably look cryptic (i.e., Harmonization) or even totally unclear (i.e., Unsponsored). Still, even at this early stage, I believe it is possible — and important — to give you a taste of the generality, utility, and potential value of the framework.

Our goal in this overview is to taste the nature and value of the framework while acknowledging these yet-to-be-explained categories. I say "our" and "we" because you, the reader, will also have an active part in this overview. Together, by my asking questions and your giving answers, we will examine the five dimensions of the framework by applying it to a concrete example.

First, I ask you to spend a few seconds selecting a standard that particularly interests you. You can use any of the standards that I presented earlier, or ones that you see or would like to see around you. You can choose the cable standard (say its short name is "Cable"), the standards for computer based characters ("ASCII"), the structure and size of credit cards ("Credit card"), tests like the Scholastic Aptitude Tests ("SAT"), or the fact that you need a tie in some restaurants ("Tie-in-a-restaurant"). Better yet, you may want to select a standard from your own setting. (You don't have to spend too much time. In talks I have given about standards, I found that the first thing that comes into your mind usually suffices.)

In any case, make sure that you have a name for the standard, preferably a short name (up to four words is best). Then, in the following paragraphs, we will use the framework together to ponder about the Level, Purpose, Effect, Sponsor, and Stage of your standard.

The Level dimension will prompt us to think about the users and producers of the standard. For example, if you chose the SAT standard, then the users are students (Level-individual) and universities (Level-organizational), and the producer, one in this case, is the Educational Testing Service (Level-organizational). Who uses your standard? Is it used by individuals, organizations, perhaps nations or the entire world? Was it developed by one of the international bodies, or perhaps by an association of companies? Perhaps it was developed by a particular person?

The Purpose dimension will prompt us to think about the aims, both intended and actual, of standards. For example, the "Tie-in-a-restaurant" standard is aimed at maintaining a respectful clientele and protecting those who want to get their money's worth in terms of ambience (Purpose-protection). What about your standard? Perhaps it was originally intended just to create vocabulary, or perhaps it was intended to protect consumers from potential harm. Some standards, and yours may be among them, were originally designed to support simplification, but later they were used to support protection.

The Effect dimension will prompt us to consider the pros and cons, the benefits and problems, and the payoffs and tradeoffs that standards have. If you chose the Cable standard, then a payoff would be the diverse channels that we can now enjoy (Effect positive) and the tradeoff would be the monopolistic system that the cable industry operates in (Effect-negative). What about your standard? For example, it may currently have positive Effects on one organization, but long-term negative, and perhaps even destructive Effects on another organization. Or just the opposite; it may have negative Effects now, but constructive Effects in the future. We may also find that we basically know nothing about the Effects of your standard.

The Sponsor dimension will prompt us to consider the origin of the standard. In the case of the credit card size, the sponsor is the International Standards Organization (Sponsor-multisponsor). Who developed your standard? Can you identify it? Was it a single entity that is making lots of money off it? Or perhaps a not-for-profit coalition of many organizations? Is it a standard with a punishment attached to it, or just a recommendation?

The Stage dimension will prompt us to think about the process of making the standard. For example, the ASCII standard is well established (Stage-existing), although there is some discussion about extending ASCII to include non-roman languages (like Arabic and Hebrew). What about your standard? Does it already exist? Is it widely used by many people? Perhaps its use is already declining, as its negative Effects overcome its positive Effects?

The above brief mental experiment should give you a taste of the framework's working. In essence, the five dimensions act as mental prisms. Like real prisms, which are used to break down and analyze light into its basic colors, the dimensions can be used to break down and analyze an object into its basic components. The object in question can be a particular standard, a setting, a view, or some other target of analysis that involves standards.

In some cases, with certain objects, several categories or even whole dimensions will not be applicable. Yet by having all five dimensions in our mental arsenal, we equip ourselves with a general tool. The price of this generality is the lack of applicability of some of the dimensions to some cases. This may explain why, in the above mental experiment, you might have found that particular dimensions did not relate to your selected standard.

ON DESIGNING THE FRAMEWORK

Building on the dimensions discussed in previous sections, this section presents reflections on the process of developing the framework. This section attempts to answer several questions: What dimensions did not make it into the framework? Why were certain names chosen for the categories? And what is difficult about the current framework?

This section is not intended as a full-fledged evaluation of the framework, rather, as the starting point toward one. My intention here is to cover a lot of ground in relatively little space, so that future work can build upon this ground.

Which Dimensions Did Not Make it into the Framework?

Selecting the five dimensions, Level, Purpose, Effect, Sponsor, and Stage, was not easy, because the literature suggested many other possible dimensions to standards. This section offers a kind of obituary for those not included, that is, short, simplified descriptions aimed at encouraging further research.

The dimension that almost made it, called "Form," concerned the embodiment of standards. Form included, as usual, five categories-definitions, specifications, processes, certifications, and meta-standards-in near final format (using the suffix "-s"). Beyond the format, preliminary definitions were developed for the categories; yet as the logic, arguments, and examples for the Form dimension were being outlined, distinctions between the categories faded. Further, the different Forms of standards that were distinct in the industrial age seemed to lose distinctiveness in the knowledge age, when a single standard often takes multiple Forms. After the Form dimension died, one of its categories (meta-standards) was incorporated into the Purpose dimension (under harmonization).

Another dimension that did not make it into the framework (Sivan, 1993b) was "Domain," which included the categories of business, knowledge, and technology. This dimension, which captured the three cultures presented earlier, was intended to push toward a focused analysis of the cultures that motivate the producers and users of standards. Yet as work progressed, those cultures seemed more the background for the whole framework, rather than a separate dimension. Using Domain as an analytic dimension produced interesting results, although these were related more to the emerging nature of the knowledge age than to standards.

Beyond "Form" and "Domain," several other candidate dimensions appeared in the literature. Apart from one in Verman's (1973) work, all of them found their way into the framework either as actual dimensions or as parts of dimensions.

Verman's "Subject" dimension (pp. 48-58) includes the categories of engineering, transport, housing/building, food, agriculture, forestry, textile, chemicals, commerce, science, and education. Verman considered this a partial list and said that further disciplines would need to be incorporated over time. The Subject dimension did not make it into my framework because it seemed too specific to the working of National Standardization Bodies, which are the focus of Verman's book. Further, because in the knowledge age the interdisciplinary links are as important as the disciplines themselves, there is no particular point in highlighting the disciplines of the Subject dimension.

Other than these three dimensions, several concepts did not find their way into the framework, such as quality, market, competition, monopoly, consensus, control, and voluntary. They might have provided the basis for a dimension tentatively called "Principles" (because all could be considered principles of standards), but on reflection, most of these concepts could be included as parts of dimensions already in place (e.g., quality under the valuation category of Purpose, or monopoly under the unisponsored category of Sponsor). Without a potentially explicit logic for these tentative categories, the Principles dimension was not pursued.

Why These Names for the Categories?

After all sources were read and all considerations made, and after the dimensions were settled, came the final process of naming the categories. The process of name-smithing was really one of fine-tuning, in which alternative considerations, arguments, and competing agendas were hammered into the specific names. To arrive at the names, several guidelines (presented below) were used, ranging from assuring the substantive soundness of a name to enhancing its aesthetic appeal in relation to the rest of the framework.

These guidelines were used in naming the categories of all five dimensions. By design, they blur the differences between substantive and aesthetic.

- As much as possible, prefer names from the sources, especially names cited repeatedly; a case in point is Verman's (1973) Levels (individual, organizational, associational, national, and multinational), which, with only slight differences, appear in more than ten sources.
- Prefer general over specific names (i.e., organizational rather than industrial), to support use of the framework in diverse settings.
- Prefer names in their common meaning and do not invent new meaning for words; prefer extending common meanings, rather than selecting esoteric meanings found only in the dictionary. (The word "harmonization," which is used to mean harmonization of national standards, was chosen to describe

harmonization of any standards, in contrast to another possibility, the word "reconcilement," which has almost the same meaning).

- When necessary, create new names; when there is a reason, use a lesser known word (e.g., harmonization) or invent a word (e.g., unsponsored).
- Prefer simple names. Give a high priority to the usefulness of the framework; thus, as a rule of thumb, the category names should be known words that a lay person can readily understand.
- Do not use the same name for two different categories, in order to prevent confusion; for example, the same word could have been used for the unknown Effect, the devoid Sponsor, and the missing Stage, all of which are in some senses similar.
- Do not use nonstandard categories; all categories should look the same. This led to selecting one word, rather than two words or a hyphenated expression (e.g., multinational rather than multi-national).
- Prefer names with different initial letters for categories in the same dimension. Although this guideline could not be satisfied for all dimensions, it helped produce interesting names.
- Strive for structural similarities (e.g., using the same prefix or suffix) in each dimension; similar categories contribute to the usefulness of the framework. For all dimensions suitable suffixes were found: Level ends with "-al," Purpose ends with "-tion," Effect with "-tive" (except unknown), Sponsor with "-d," and Stage with "-ing."

In selecting the final names, substantive soundness of the category names preceded aesthetic attributes. If a choice had to be made between two substantively sound words, the more aesthetic one could be selected. All but one name satisfied both substantive and aesthetic considerations (the exception being the less pleasing "unknown," as opposed to the more pleasing "none," which, however, terminates with an "e" and did not satisfy the logic criterion).

What Is Difficult About the Current Framework?

Although the most appropriate category names were selected to support the use of the framework, the names of some categories will undoubtedly cause problems, raise questions, or, worse, be misunderstood. This forecast is extrapolated from responses both to previous versions and to the current framework.

The Level dimension will probably present only minor difficulties, because it has a powerful intuitive and lucid logic. The individual and the associational categories may give rise to questions, which can be answered quickly by revisiting the definitions and examples.

The Purpose dimension and its categories are a different story. Of the five dimensions, this one may prove to be the most difficult. It lacks an ordinal logic. It includes unrelated categories, at least to the eyes of a typical novice observer. It uses a relatively new word (harmonization). And it includes the emotion-laden "valuation" category. These problems are likely to subside after several uses of the framework.

The Effect dimension is the least problematic one. Its symmetric structure provides a clear logic. There are only two potential problems: the distinction between the two mild and stronger categories (positive and constructive, negative and destructive) and the meaning of the unknown category. Both can be resolved by reading the definitions and examples.

The Sponsor dimension is expected to present a challenge. After Purpose, it seems the most problematic. The categories have no intuitive meaning: the use of "devoid" and "mandated" may not be clear in the context of standards, yet, paradoxically, may help users, who may find assigning new meanings to new words easier than assigning new meanings to previously known words. The three different prefixes ("non-," "uni-," and "multi-") to the same word ("-sponsored") may also be helpful.

Last, the Stage dimension should present only minimal problems, since its logic is obvious. The category names, particularly "existing" and "dying," may prompt questions, even objections, but novice users should otherwise have no difficulty relating to this dimension.

One potential problem for all the dimensions is the boundaries between categories. This is especially true when several categories may be applicable. For the sake of generating insights, clear distinctions between categories are not needed; indeed, some fuzziness may encourage creative use of the framework.

Despite these potential problems, the logic and aesthetics of the dimensions should make them easy to understand and therefore easy to use.

ON THE FRAMEWORK IN GENERAL

In this section, I will expand on judging the value of the framework, on improving the framework, and on other complementary insights.

Judging the Value of the Framework

The rationale for the framework for standards was developed in terms of its logic, precedents in the literature, and the ability to generate general insights. But more broadly, how does one tell whether a framework is "right?" What would validate it? A framework is not a theory. It does not directly advance claims that can be tested, like "smoking causes cancer" or "gravitational fields deflect light." Instead, a framework is a classification system, its soundness not a matter of truth (because it makes no claims) but of organizational usefulness—how completely and clearly does it classify? The Periodic Table of Elements, for instance, is not true or false (although individual atomic weights of particular elements may be true or false) but, rather, complete, clear and illuminating.

At this point, I want to argue that the long-term soundness of the framework depends on a demonstrably good track record in delivering insights. Whether the framework "delivers" in this sense can be found through an experiment that addresses not the truth (because, again, the framework does not claim truth) but the usefulness of the framework. Such research is beyond the scope of this work, but imagining such a study is a useful mental experiment.

A conceptual experiment-a skeleton structure here, not a fully designed experiment-if conducted properly, might reveal how the framework should be judged.

Conceptual Three-Phase Experiment for Proving the Value of the Framework

Phase 1: In this phase of the experiment, two groups of ten people (the "subjects") will watch a 30-minute video called "Introduction to Standards." The first, or "framework," group, is also exposed to a 5-minute video that presents the framework. The second, or "control," group, is not exposed to this video.

Phase 2: In this phase, the subjects in both groups are asked to "use what they have learned about standards" to deal with several tasks (e.g., analyzing settings, analyzing standards, analyzing views, selecting standards, designing standards). They are also told that they will be asked to report on "where, how, and how well they have used the concept of standards."

Phase 3: In the last phase, after dealing with the tasks for a few hours, the subjects' responses are collected and analyzed by "blind" evaluators (who do not know the original grouping of the subjects). The evaluators are asked to assess the quality and effect of the video, on the basis of the subjects' reports about their use of the concept of standards.

Potential Results

If the "framework" group does better than the "control" group (as assessed by our blind evaluators who will examine the quality and effect of the video for each subject), then, yes, the framework helped people to apply the concept of standards. This result would not mean that this is the ultimate framework; it would simply mean that the current version had positive effects.

But if both the "framework" and the "control" groups were to have the same result (again, as assessed by blind evaluators), then, no, the framework did not help then to apply the concepts of standards. On the assumption that there were no flaws in the experiment, the particular framework did not work, but this result would not mean that other frameworks would not work.

To turn this conceptual experiment into a real one would require careful consideration of the different parameters (e.g., number of subjects per group, length of video, number and kinds of tasks, length of time spent performing the tasks, and the training method of the "blind" evaluators). If conducted correctly, the experiment would gauge the value of the framework.

In conclusion, the mere fact that an experiment to test the framework can be suggested means that this is indeed a "scientific" framework, that, in theory, the value of the framework can be gauged *empirically*. The experiment is a reminder that the value of the framework is its real-world ability to generate insights about issues.

Lastly, the three phases of the conceptual experiment can be carried out by those who wish to try the framework themselves. They can study the framework, apply it to several tasks, and then reflect on its usefulness and value.

Can the framework be used in its current format? Probably yes. Although it may call for some experimentation, it is possible to use the framework in various ways for various settings. The interested reader may re-examine how earlier in this chapter the framework has been used: (i) to analyze a setting that involves standards, (ii) to analyze a particular standard, (iii) to analyze a view, (iv) to select a standard, and (v) to design a standard. These uses may hint at the general generative power of the framework.

The framework can be used by both novice and expert thinkers about standards. Novice thinkers can use the dimensions as an analytic checklist, in which the categories of each dimension prompt analytic questions about the particular issue. Expert thinkers, more familiar with standards and thus with the spirit of the dimensions, will probably use the analytic matrices where two or three dimensions highlight an issue.

Improving the Framework

Unfortunately, the current framework has one major flaw, a flaw so big it may hamper the long-term usefulness of the framework. As it currently stands, the framework, which I alone developed, lacks an established Sponsor, one that could worry about its long-term survival and prosperity, update it as needed, and maintain its integrity. Such a Sponsor would probably test the framework more extensively than a single person can and build on it in various ways for the benefit of all users and producers of standards.

As with any other standard, before people can use the framework, they need to trust it. They need to count on the Sponsor of the standard to be reliable a few years down the road. A Unisponsored standard is inherently weaker than a Multisponsored one, because users do not trust it. Lack of trust discourages adoption, which, in turn, can bring the standard into the Stages of decline and, ultimately, dying. Seeking reliable institutional sponsorship is far beyond the scope of the initial development of the framework. Potential sponsors may include national bodies, such as ANSI, or, preferably, international bodies, such as the ISO. For long-term survival and prosperity — *the framework needs an institutional Sponsor.*

Beyond the need for a Sponsor, is this the best framework? Absolutely not. Although it represents the results of a fair, even exhaustive, effort, there is always room for improvement. The framework must be improved to support the evolving uses and circumstances of the knowledge age.

There are several instances where fine-tuning the names of the categories may contribute to the use of the framework. Fine-tuning might also involve changing the order of the categories and expanding their definitions. A more substantive improvement may be needed when new dimensions become more salient (e.g., the Domain dimension, which did not make it into the current framework, might become more important with further entry into the knowledge age).

Additional models could be developed that might examine uses of the framework in different fields and different languages. A typical model use can follow the exploratory section presented earlier in this work. Single dimensions can be used as analytic checklists, and two or three dimensions can be used as analytic matrices to analyze settings, standards, and views, and to select or even design standards.

Another potential aid that could enhance the usefulness of the framework would be a model, called "the dimensions and other concepts," that might even take the form of a poster on which the five dimensions would be linked to other concepts that relate to the field of standards (e.g., monopoly, competition, consensus, voluntary, and anticipatory, among others). This model might be especially useful to the standards community.

In summary, the framework should evolve to accommodate changing needs, diverse settings and users, and related concepts. Thus, in more general terms, the *framework needs to be updated constantly*. But a caveat about updates:

- All updates must maintain the flexibility of the framework in order to accommodate future changes; a particular virtue of the current framework, which stems from the independence of the dimensions, is its ability to accommodate change. Because the dimensions are not directly related to one another, dimensions can be added, modified, even deleted without endangering the integrity of the entire framework. This "independence that allows flexibility" should be guarded.
- Even more important, updates should be grouped and not cause frequent changes. Because users need to assimilate the changes, constant and frequent changes might harm the framework's most important role, which is to facilitate dialogue about standards by creating a common vocabulary. If the language is not stable, people will not be able to use it. Thus, *the framework needs to be both flexible and stable*.

Complementary Insights Beyond the Framework

In the course of mapping the general land of standards, beyond the general framework of standards, other insights were gained. Five insights worth noting here:

(i) There is a lack of general frameworks for standards.

Reexamining the use of the sources, I was struck by the lack of any attempt to talk about standards in general. Aside from Verman's (1973) book *Standardization: A New Discipline*, and the OTA (1992) report Global Standards, all other sources took a more focused and limited look at standards. Most analyzed and compared a few cases, and some analyzed what may be called one or two dimensions. This finding largely confirms what can be described as "omnipresence."

(ii) In the knowledge age, as opposed to the industrial age, standards will play a greater role.

Understanding standards requires understanding their roles in the industrial age as well as their potential role in the knowledge age. This understanding can come from examining the culture and sphere images presented above. Although simplistic, the shift from the industrial age cultures of business and technology that operate in a sphere of standards to the knowledge age cultures of business, technology, and knowledge that operate in a bigger sphere of standards serves as a reminder of the growing roles of standards in the knowledge age.

(iii) The framework's lack of exclusiveness may look like a bug but it is a feature.

An examination of the framework reveals that the five dimensions can be related to other concepts besides standards. For example, the dimensions of the framework could be dimensions of "cultures," of "forces," of "computers," or of almost any other sufficiently large concept. Only small, concrete objects cannot use the dimensions (e.g., lamps, pens). I tried to apply the dimensions as dimensions of forces and of cultures and so on, and the results of this limited application revealed that while the dimensions seemed to work well when the concept was close to standards (i.e., force, quality), they did not work as well when the concept was further from standards (i.e., computers, markets). Thus, in spite of appearing to be a bug, the lack of exclusiveness of the framework is really a feature that reflects the framework's generality. The framework was, after all, designed to be used across many settings. Still, this potential flaw may require additional, more specific standards-based terminology.

(iv) Aside from standards themselves, producing them has important benefits.
Arriving at standards creates informal opportunities to discuss, understand, and partner with interested parties. The need to arrive at specific ways to phrase, measure, and define the standards forces those with different mindsets to communicate and bridge their differences. These side benefits, which were not explicitly part of the framework, may be incorporated into future versions.

(v) Producing standards demands skilled producers.
Good standards may serve as critical leverage points for solving greater problems, but arriving at the right standards at the right time and at the right place is far from trivial. To produce a standard in a particular field requires, beyond extensive knowledge of the field, the ability to lead and follow, negotiate internal and external competing agendas, balance present and future needs, and, perhaps the most challenging aspect of standards production, the ability to juggle the skills of abstraction as well as simplification

REFERENCES

Batik, A. L. (1992). *The engineering standards: A most useful tool.* Book Master/El Rancho.

Brady, K., Taylor, E., & Willwerth, J. (1990, October 8). Suffer the little. *Time,* 39-48.

Dertouzos, M. L. (Ed.). (1989). *Made in America: Regaining the productive edge.* Cambridge, Mass: MIT Press.

Drucker, P. F. (1980). *Managing in turbulent times.* New York: Harper & Row.

Gaillard, J. (1934). *Industrial standardization: Its principles and application.* New York: H. W. Wilson Company.

International Organization of Standardization [ISO] (1990). *A vision for the future: Standards needs for emerging technologies.* Geneva, Switzerland: ISO.

Kent, W. (1978). *Data and reality: Basic assumptions in data processing reconsidered.* New York: North-Holland Pub. Co.

Lemonick, M. D. (1992, Fall). Tomorrow's lesson: Learn or perish. *Time, Special issue. Beyond the year 2000,* 140, 59-60.

Naisbitt, J. (1982). *Megatrends: Ten new directions transforming our lives.* New York: Warner Books.

Naisbitt, J. (1990). *Megatrends 2000: The new directions for the 1990's.* New York: Morrow.

Office of Technology Assessment, United States Congress [OTA] (1992). *Global standards: Building blocks for the future* (TCT-512). Washington, D.C.: Congress of the United States, Office of Technology Assessment.

Perkins, D. N. (1986). *Knowledge as design*. Hillsdale, N.J.: L. Erlbaum Associates.

Sanders, T. R. B. (Ed.). (1972). *The aims and principles of standardization*. Geneva, Switzerland: International Standards Organization [ISO].

Senge, P. M. (1990). *The fifth discipline: The art and practice of the learning organization*. New York: Doubleday/Currency.

Sivan, Y. Y. (1993a). *Project Y: Toward standards that link business, education, and technology: The case of a university computer center*. Unpublished doctoral dissertation : Harvard Graduate School of Education.

Sivan, Y. Y. (1993b, Summer). The pandora's box of standards for education. *Technos*, 2, 19-21.

Toffler, A. (1971). *Future shock*. New York: Bantam Books.

Toffler, A. (Ed.). (1974). *Learning for tomorrow: The role of the future in education*. New York: Vintage Books.

Toffler, A. (1981). *The third wave*. New York: Bantam Books.

Toffler, A. (1990). *Powershift: Knowledge, wealth, and violence at the edge of the 21st century*. New York: Bantam Books.

Verman, L. C. (1973). *Standardization: A new discipline*. Hamden, Conn.: Archon Books.

Chapter II

Consensus Versus Speed

Roy Rada
University of Maryland, Baltimore County

INTRODUCTION

One standards organization takes years to reach consensus, while another standards organization takes months (Rada, 1995a). Changes are occurring in the most famous and internationally powerful standards organization, the International Organization for Standardization (ISO), to speed its consensus process. However, other organizations already combine some consensus with exceptional speed and may have greater impact on the standards world.

The tension between consensus and speed is inevitable. New technologies support communication and decision-making, and these technologies should be used to narrow the gap between standards' processes that are strong in consensus and those that are fast. Consensus by those who bless the development of standards does not necessarily imply consensus in the community of those who use it. Both kinds of consensus are important, and information technology tools for standards development and dissemination can help achieve consensus.

This chapter will examine the attempts by standards organizations to increase speed without sacrificing consensus. The analysis reveals that organizations are cooperating in new ways that should increase the speed with which standards documents arise. Different communities of consensus result. The traditional standards development organizations continue to prevent the free flow of standards documents on the Internet. These documents should be free.

BACKGROUND

Natural language is a standard within a culture and usually evolves continually. Speed versus consensus is reflected in the arguments about whether popular, new terms are unacceptable jargon or whether they belong to proper language. A different type of common language comes from units of measurement. One of the earliest types of measurement concerned that of length. Length measurements were usually based on parts of the body. The first documented example is the Egyptian cubit that was derived from the length of the arm from the elbow to the outstretched fingertips. By 2500 BC this had been standardized in a royal master cubit made of black marble. In England units of measurement were not properly standardized until the 13th century, though variations and abuses continued until long after that (Dictionary, 1999). Speed of some standardization efforts can be measured in millennia.

This chapter focuses on international standardization of information technology. The issues here are rather different from those experienced in the evolution of natural language standards. The analogy to units of measurement would not be far fetched in terms of the problems of consensus but the issues of speed are different. Information technology is changing at a rapid rate and requirements for standardization are special.

Consensus but not Speed

ISO is an independent organization for fostering international agreement on standards with a view to expanding international trade. ISO formally consists of national representatives only. To develop standards, ISO relies on a large number of volunteers who participate in more than one hundred Technical Committees. Each Technical Committee in turn relies on Working Groups. Working Groups develop recommendations as working drafts, then committee drafts, then draft international standards, and finally international standards (ISO, 1999). At each stage, people follow an arduous cycle of corresponding with one another, meeting face-to-face, and formal balloting. When each Working Group feels each text is good enough, the draft is forwarded through channels to the national standards bodies. Each member country places one vote at each stage. The ISO process from first correspondence to a published international standard typically takes years.

The scope of ISO standards is not limited to any particular area. ISO covers almost all standardization fields. If standards for screw threads are reconsidered by standards developers every five years, the market is satisfied that the standard is adequately maintained. For information technology standards such maintenance is too slow and too distant from market concerns. This means a process right for screws (there is in fact an entire Technical Committee of ISO devoted to screw threads) but not computing.

ISO has agreements with the national standards organizations that require those standards organizations to accept ISO's international standards. However, neither ISO nor the national standards organizations are in a position to mandate that industry uses the ISO standards. For instance, some years ago ISO initiated the Open Systems Interconnection (OSI) standard. OSI covers computer networking, and the American government initially felt that following OSI should be mandatory in government procurements of computer networking equipment. However, the success of the Internet, that did not specifically follow the OSI model, led the government to realize that adherence to the ISO standard approach was not cost-effective.

In partial recognition of the special character of information technology, ISO joined forces in 1987 with the International Electrotechnical Commission and created a Joint Technical Committee on Information Technology (called JTC1). However, JTC1 until recently had to follow the time-consuming steps and strict, copyright rules that other Technical Committees of ISO follow. In response to advances by other standards organizations, such as the Internet Engineering Task

Force, ISO has had to modify its approach and in particular started with giving JTC1 new options.

Speed but not Consensus

The purpose of the World Wide Web Consortium (W3C) is to support the advancement of information technology through standardization in the field of networking, graphics, and user interfaces that support the continued evolution of the Web. W3C was founded in October 1994. It is an international industry consortium, jointly hosted by Massachusetts Institute of Technology in the United States; Institut National de Recherche en Informatique et Automatique in France; and Keio University in Japan (World, 1999).

W3C consortium members pay fees. No individuals can join. Full members pay $50,000 per year for membership. Affiliate members pay $5,000 per year for membership. All members must commit to three years of dues on first joining. W3C has over 230 members. These members include companies with 100,000+ employees, such as Boeing, British Telecom, Citibank, Matsushita, Siemens-Nixdorf, Sun Microsystems, and Xerox and software companies, like Microsoft and Netscape, whose employee base is relatively small but whose investment in the development of proposals for new Web standards is substantial. The members not only contribute their membership dues for staffing of W3C, but the members provide proposals and implementations from their large work force. These are then studied by the membership of W3C, and agreement is sought as to which proposals and implementations deserve most to be supported as standards for the Web.

Speedy Process

A proposal for a new standard goes through the following process. A proposal must be sponsored by a member organization and must include a complete draft specification of interfaces and functionality. The sponsoring organization should present the proposal to the Consortium Director for formal consideration, and designate an individual who will act as an Architect for the specification. The Director then ensures that the proposal is circulated to the W3C members, and a one-month trial review is begun. At the end of the trial review, each member is asked to give the Director a Yes/No vote. The Director considers the votes, and makes an informed decision as to whether to proceed with the technical review.

If the decision is to proceed, then a three-month technical review begins. A Web posting and an electronic mailing list are created at MIT, INRIA and Keio for discussion; participation on this list is open to all member organizations, without constraint. As issues are raised during the review, it is expected that participants will vote (formally or informally) on how to resolve the issues. The Architect considers the votes, and makes an informed decision. Meetings may be called, as deemed necessary by the participants or the Director.

At the end of the technical review period, each member is asked to give the Director a vote. The Director will consider the votes, and make an informed decision as to whether to proceed to the next stage of the process or not.

If the specification is accepted, then a three-month public review begins, accompanied by a posting on the Web of the specification. In parallel with the public review, a 'proof of concept' implementation of the specification is undertaken. Proof of concept typically requires a complete, public, portable implementation.

When public review has ended, another vote of members is taken. The Director will consider the votes, and make an informed decision as to whether to accept the specification or not.

Once a standard has been approved, the Web posting and mailing list will continue to be used to discuss possible clarifications or additions to the standard. In general, all materials included in Consortium-sponsored distributions will be copyrighted, and will contain permission notices granting unrestricted use and redistribution of the materials provided that copyrights and permission notices are left intact.

The entire process from start to beginning of standardization in W3C can occur in a matter of weeks. Months are typically required but not usually years. This contrasts sharply with the JTC1 process on a given proposal that typically took years to progress from beginning to end. On the matter of speed, W3C is to JTC1 like the rabbit is to the turtle.

Consensus

W3C allows its members to voice their opinions. Also the public can criticize any W3C proposal or implementation during the public review phase. However, almost every time a decision has to be made, it is left entirely to the Director. The Director is told to consider the votes but then to decide in whatever way the Director feels is best.

Any organization can join W3C, whereas only countries can join JTC1. The members of W3C are largely those with a vested interest in selling the technology. On the other hand, these same organizations are likely to be influential in their country's voice to JTC1. Why should a company with 40 employees have the same vote as a company with 400,000 employees, or a country with 10 million people have the same vote as a country with 1 billion people? A formula to decide how much weight to give to each vote might be so cumbersome that it would be better reduced to having one good Director who decides.

The W3C standards process has been the source of some controversy. Claims have been made that one member has manipulated or completely controlled the process for a given standard. Some members have attempted to get Architects fired because they had accepted consulting work from other members. The consortium does not bring such controversies to the public.

So there is speed, but is there consensus? W3C has brought forth several standards in its short life and is poised to produce many more. Some say these standards represent the consensus view of the world's most concerned participants and help bring the most appropriate technology to the people. Some say W3C is rubber-stamping the products of major vendors.

Speed and Consensus

The difference between the ISO standards processes and others is particularly pronounced when considering the process in which standards are made for the

Internet. The key organization in the Internet standardization process is the 'Internet Engineering Task Force' (IETF) which does not operate under the umbrella of ISO.

The IETF is a large, open, international community of network designers, operators, vendors, and researchers concerned with the evolution of the Internet architecture and the smooth operation of the Internet. It is open to any interested individual (Internet, 1999). The American government through the Corporation for National Research Initiatives funds the IETF Secretariat.

The IETF standards process begins with someone expressing an interest and then moves to the creation of a Working Group of volunteers. These volunteers communicate extensively by email and make draft documents publicly accessible. While a Working Group may have meetings, no final decisions are made in face-to-face meetings but are made over email.

An IETF Standard is preceded by a Draft Standard that in turn is preceded by a Proposed Standard. Proposed Standard status can be reached arbitrarily quickly. The Proposed Standard must have demonstrated utility and be publicly available for six months before it passes to Draft Standard status. A Draft Standard must have multiple, independent implementations. With relatively little bureaucracy, a document may pass from one stage to another and reach permanent archive status as an Internet Standard.

IETF standards are different from ISO standards and are not part of an official, government standardization effort. IETF standards are also different from 'de facto' standards. A 'de facto' standard is not publicly developed but is widely used, such as MS-DOS (Rada et al, 1994). IETF Standards are publicly developed and are often widely used.

Some would say that the early success of IETF was based, in part, on its having been not seen as important commercially. Now that companies want to influence the Internet, the IETF may have to introduce further hierarchical controls to assure consensus. These additional controls would, in turn, slow the standards process.

What to do?

IETF and ISO are different in more ways than sketched here. IETF expects implemented standards, whereas ISO accepts hypothetical standards. The ISO standards process has traditionally been based on the circulation of paper drafts of standards and face-to-face meetings to discuss revisions to these drafts. This cycle of circulating paper drafts and traveling to meetings to discuss changes often takes years. The IETF on the other hand has reduced the need for many, long, face-to-face meetings through the frequent exchange of drafts via the Internet and through extensive email discussion of the content. The open, digital libraries that the IETF maintains help volunteers speed their standards development and still maintain consensus. The success of the IETF approach speaks for itself.

Shortcuts are being introduced into the formal process of standards developers. This chapter will examine the controversies that arise from these shortcuts. The Publicly Accessible Specifications process of ISO will be critiqued relative to the Sun Corporation submission of Java. Then the alliance between the World Wide Web Consortium and ISO will be assessed. In both cases, the consensus process within the organization that is proposing standards to ISO/JTC1 is not the same as the consensus process in ISO.

One solution is to acknowledge various levels of documents in terms of their achieved degree of consensus. This approach is illustrated in the new CEN Workshop, which allows for documents in various stages of development and consensus to be publicly endorsed and accessible. A problem remains to get these documents to people widely. One solution is free documents on the Internet, as long done by IETF.

CONTROVERSY

One way for ISO to increase speed is to take standards developed quickly by some other organization and convert those into ISO standards. This raises serious questions of whether consensus has been adequately served. Next examined are two cases of such shortcuts implemented by ISO and comments on their appropriateness.

Shortcuts

The General Assembly of ISO, in extraordinary session on November 4, 1993 in Geneva, endorsed a package of proposals for restructuring the organization in order to meet the challenge of producing standards faster (Standardization, 1994). One reflection of this restructuring was an agreement with JTC1 to test a new method of involving consortia in quick approval of their standards (Smith, 1999).

Other options also exist for speeding the acceptance of a standard in JTC1 — one such option is called the 'fast track'. In the fast track, existing members of JTC1 can submit a document for approval as a standard directly to the top JTC1 level. Existing members of JTC1 are predominantly national standards bodies. These national standards bodies typically mirror ISO methods of working. In other words, this fast track to JTC1 avoids the detail work of JTC1 but requires the similar detail work of a national standards body.

The PAS Process

The intention of the Publicly Available Specifications (PAS) shortcut process is to attract consortia to submit their proposed standards to ISO. However, the rules specify that any organization can attempt to participate whether it represents a consortium or not. The shortcut is described as the transposition of PAS into international standards (ISO/IEC, 1995).

The PAS process is similar to the fast track process, except it introduces a new category of organization that can submit a specification directly to JTC1 for approval. These are PAS Submitters. Information technology solutions that have been developed outside the formal standards making community may through PAS be given the status of ISO standards.

In 1997 Sun Microsystems Incorporated (Sun) submitted a PAS proposal for Java. Sun did this, of course, because the blessing of ISO brings further market share to the blessed. Should a single company make an ISO standard? An exciting battle was fought on the PAS field between the allies of Sun and their enemies. Sun won the battle to become recognized as a PAS Submitter but as of early 1999, Java was still not an ISO standard.

Any organization with a document that may be considered suitable for transposition into an international standard is a PAS Originator. There are no fundamental restrictions as to what form the organization should take. A PAS Originator must apply to JTC1 for recognition as a PAS Submitter. This application must answer certain questions about characteristics of the PAS Submitter and must describe in brief the first PAS that the Originator intends to submit. The national members of JTC1 then have three months in which to vote on whether to accept or reject the application to be a PAS Submitter.

The criteria that an organization must satisfy before being accepted as a PAS Submitter are grouped into three categories: cooperative stance, integrity, and intellectual property rights. Cooperative stance concerns the organization's history of working agreements with other organizations for the purposes of standardization, its plans for maintaining its specifications, and its receptiveness to externally generated change requests. The integrity of the business is related to its longevity and reputation. The intellectual property rights criteria address the organization's preparedness to grant patent, copyright, distribution, or trademark permissions to other organizations.

The main concern about the PAS Submitter should be broadly speaking whether or not the submitter develops its standard in an open way. This means that input from experts worldwide is somehow respected. Only after achieving PAS Submitter status can an organization submit a PAS for possible transposition to an international standard. Then the focus is on the content of the specification itself and no longer on the process by which the specification was produced. The PAS itself must satisfy criteria for quality, consensus, and alignment. The member bodies of JTC1 vote on whether the PAS meets the criteria.

Prior to 1997 only a few organizations had become PAS Submitters and they were all consortia. For instance, one PAS Submitter is the Infrared Data Association, a consortium of over 150 companies. X/Open was also approved by JTC1 as a PAS Submitter. Every National Standards Body voted in favor of X/Open's application. According to X/Open members, the benefit to X/Open is that X/Open gets a direct route into the formal standards process. Prior to achieving PAS Submitter status, the only practical route to ISO available to X/Open was via some other standards organization that itself reported to JTC1. The aim for X/Open is to reduce the cost and risk of change, when it takes its specifications forward to become ISO standards.

The Sun Case

Given that the guidelines for moving a PAS to an international standard have existed since 1994, relatively little attention had been given to PAS until Sun requested to become a PAS Submitter for Java. The Sun submission created an enormous amount of interest both because of the precedent it would set in having a commercial company be a PAS Submitter and because of the large market influence of Java. Sun requested PAS Submitter status in March 1997.

Sun explained that for several years it has been sending representatives to various JTC1 working meetings and that it will continue to do the same. About maintenance, Sun said it is committed to evolving the Java platform in response to market conditions. Sun owns and intends to continue to exclusively own its Java trademarks, such as the names 'Java', 'Java compatible', '100% Pure Java', and the 'cup and steam' logo.

The first JTC1 decision about whether to accept Sun as a PAS Submitter was mixed. Voting in JTC1 is done by the national bodies that are members of JTC1. Each nation has one vote, and about 30 nations are represented. The national bodies collect opinion from their constituencies. Initial opinions of some constituents of the American National Standards Institute are next described.

In defending Sun's application, the US Department of Defense said, "This particular issue is a watershed issue for ISO...a 'yes' vote is very important .... JTC1 demonstrates to the world its intent to be relevant to technology .... A 'no' vote would send the wrong message that others need not apply."

A May 1997 letter from Brad Silverberg, Senior Vice President of Microsoft to JTC1 included the following: "Sun wants the benefit of an ISO standard, but is unwilling to part with even the minimum intellectual property rights or to commit to an open process for developing the technology in the future. ... Microsoft can only conclude that Sun is unwilling to commit to an open process, yet is intent on achieving the appearance of such a process that ISO status for Java would provide." Sun and Microsoft were battling over the fair way to standardize.

From other countries the input to the member bodies was also varied. The Austrians voted in favor of the Sun application with the following rationale: "We believe Sun's proposals-to-be should be considered on their individual merit, and would like to see them joining the ISO/IEC framework. The PAS process has been in place for a while, but we are not aware of any document generated through this mechanism. Therefore Sun's case will set a precedent anyway, and one might consider it about time that somebody actually tries to use the PAS process as it was intended."

After many debates, JTC1 decided to neither approve nor disapprove Sun's first request to become a PAS Submitter. Instead, JTC1 invited Sun to respond to the criticisms that its proposal had stimulated and that JTC1 would then reconsider the proposal with this additional information. In the repeat consideration, only the U.S.A. and China voted against the PAS Submitter proposal of Sun, and Sun was approved as a PAS Submitter in 1997.

The Sun case clearly shortcuts the normal consensus process in favor of speed. No other person or organization outside Sun is able in an open way to determine the direction of the standard. Sun retains absolute control.

Having succeeded in becoming a PAS Submitter in 1997, Sun and ISO have not substantively progressed in converting Java into an official ISO standard. This does not seem to have stopped the market success of Java.

Alliances

The World Wide Web Consortium (W3C) and its relationships to JTC1 illustrate another approach to speed and consensus. By what mechanism have W3C and JTC1 tried to link, and what is the relationship between speed and consensus?

W3C and JTC1

What relationship would one expect to find between JTC1 and W3C? JTC1 could be competing with W3C. JTC1 has committees assigned to the same topics as those on which W3C works. However, as noted earlier, JTC1 is too slow in its own standards development operations to compete successfully with fast organizations like W3C.

Another difference between W3C and JTC1 concerns revenue sources. JTC1 gets most of its revenue from the sale of standards documents. Thus JTC1 has historically been very jealous of its copyrights and completely unwilling to allow standards documents to be freely available. W3C gets its revenues from fee-paying members and distributes the standards through the Web for free.

In 1998 a draft agreement was approved between Subcommittee 24 of JTC1 (Subcommittee, 1999) and W3C that allowed JTC1 to accept W3C standards as JTC1 standards. The draft JTC1-W3C agreement says that both JTC1 and W3C follow an open consensus process and try to produce standards on the same subject.

The draft agreement says that both organizations desire to harmonize their procedures. Initial technical development work is done primarily in W3C with provision for JTC1 participation. Independent assessment is done in JTC1 with provision for W3C participation. Both organizations desire that W3C standards be adopted as ISO and IEC standards as quickly as is feasible. Technical changes to a specification once ISO/IEC processing has been initiated is only allowed when clarification is necessary or when a serious technical flaw is found that would render the specification unusable

All documents developed under this W3C-JTC1 agreement, including the final ISO/IEC standards, will be made available through ISO/IEC sources and also through normal W3C sources, including the open W3C Web site. In accordance with W3C procedures, the text of all working documents, as well as the final ISO/IEC standards, shall remain free of intellectual property right restrictions which would limit their open distribution following normal W3C practices.

Why the Agreement?

When a W3C standard is placed into JTC1 as a new work item, how long would JTC1 be expected to spend in considering it? Given that basically no substantive changes arc allowed, one might imagine a quick yes/no. Thus JTC1 would gain the speed it needs to remain credible in the marketplace. Likewise, W3C would gain a further endorsement of its standards that could have significance in various quarters, particularly government quarters.

JTC1 bylaws require that two-thirds of the member countries must vote in favor of a proposed standard before it can be accepted as a standard. W3C follows a different consensus process, and its Director ultimately makes all decisions (World, 1999). The W3C draft standards are openly circulated on the Web and gain input from many parties. JTC1 keeps its draft standards closed to the general public and gets consensus from a narrower group of participants. Consensus, if achieved, is not the same in the two organizations. Why would the two organizations want to cooperate as suggested by their draft agreement? Perhaps the complementary character of their consensus methods attracts them to one another.

In the draft W3C-JTC1 agreement, W3C retains copyright and the authority to distribute its version of its standard for free. JTC1 promises to make no substantive changes to any W3C standard, yet JTC1 will be allowed to sell the standard in paper form. This is a plus for wide distribution of standards. JTC1 realizes that it can gain more revenue by selling paper copies of standards that W3C gives for free from its Web site than JTC1 would gain by having no opportunity to sell those standards documents.

Imagine what might happen next in terms of JTC1 directions. An agreement such as that sketched between W3C and JTC1 could be extended by JTC1 to other consortia, such as the Object Management Group. In this way, JTC1 would become less a manager of standards development from start to finish and more of a filter on standards developed by consortia. Should consortia of fee-paying members be the fast developers of standards? Should JTC1 attempt to be a quality filter of consortia standards?

SOLUTION

ISO is providing increasingly many avenues for other organizations to fast-track, shortcut, or otherwise enter into the standards world without ISO technical experts considering the proposed standard. This approach of providing levels of entry into ISO makes sense from the speed perspective. Other standards development organizations are taking a similar approach, as evidenced by activity of the European regional standards organization discussed further in this section. This stratification of standards documents into different levels that are publicly accessible gives further opportunities for a certain kind of speed. Whether the user community will benefit from these documents depends on how widely they are adopted.

Levels

CEN is the European Committee for Standardization, and one of the three official European Standards Organizations under European law. CEN's mission (European, 1999) is to promote voluntary technical harmonization in Europe in conjunction with worldwide bodies and its partners in Europe. The European Commission provides about half the budget of the Central Secretariat of CEN and member fees pay most the remainder. CEN has established the Information Society Standardization System (ISSS) as a focal point for Information Society standardization activities. CEN/ISSS effectively has two major component parts:

- the (limited) traditional activity, standardization Technical Committees. There are ten such bodies, with activities such as bar-coding standardization, geographic information systems, road transport telematics, and healthcare informatics;
- Workshops: open, consensus-based activities whose principle objective is to produce specifications and guidance material, published in the form of CEN Workshop Agreements (CWAs).

The Technical Committees traditionally only had two deliverables:

- EN: the 'traditional' European Standard, leading to full Europe-wide implementation; also serving European regulatory needs of the 'new

approach'
- ENV: European Prestandard, serving as an experimental specification where the state-of the-art is not stable or full implementation not yet possible.

The year 1997 saw the start-up of the CEN Workshop Agreement (CWA). These agreements are developed in open workshops that are established and dissolved depending on the respective work items. Access to these workshops is also open to participants outside Europe. Physical meetings are limited since most work can be done by use of modern information technology. The development time of CWAs is relatively short; CEN put the first CWAs on the market in 1998.

On receipt of a formal proposal, the ISSS secretariat will help to support a proactive search for a balanced participation of the main interested sectors. A proposed workshop is described and is open to public comment on the Internet. A proposal may include one or more separate projects with their own 'deliverables'. Submission of a business plan forms part of the preparatory phase, specifying the resource requirement for each project and the sources of funding.

Proposals may be initiated by market or by public policy demand. Further, they may be in different stages of development when they are brought to ISSS. A project may consist of a new issue for exploration. Alternatively, it may be that a consortium or forum has started work and now needs a more appropriate environment to increase the breadth of participant expertise or to test its progress in the open market. Indeed, the development stage may have already ended and the creators seek an arena for validation and feedback.

Where consensus has been reached on a deliverable, the Workshop Secretariat shall pass as soon as possible any resulting document, after any editing, to the CEN/ISSS Secretariat for publication. The CEN/ISSS Secretariat shall ensure the preparation of the relevant cover page, and provide the document in *.pdf* format to the CEN Members.

One sees in the CEN ISSS CWA process something from successful organizations, such as IETF, both in process and deliverables. The Internet is extensively used as a communication and information sharing device to help reduce the time between decisions and to facilitate access by diverse peoples to the process.

The CEN CWA deviates from the successful IEFT process in two ways. IETF insists that standards be accompanied by working prototypes that implement the standard. ISSS does not require working prototypes. CEN retains control of the deliverables and does allow free access to the documents, whereas IETF provides free copies of its documents to whoever wants them across the Internet.

Free Documents

In early 1995 to facilitate information exchange, ISO started a web site at http:/ www.iso.ch/. This site includes the full catalog of ISO Standards and drafts and a meeting calendar. However, this ISO information service does not include the actual standards or their drafts.

Why do some standards development organizations provide standards for free and others charge significant amounts of money for them? The fees from the sale of standards and the stringent policies on sharing copies of standards may decrease the number of people who access these documents and may thus reduce the impact of the standards.

Many proprietary and vendor-consortia developed specifications are either on the Internet or are sold at lower prices than their formal cousins. The availability of specifications on the Internet facilitates their acceptance as de facto standards and rewards the development efforts of those that made the specifications. The acceptance of the specification is the major objective and is more rewarding to those responsible for developing the content than the monetary sale of individual copies of the standard.

ISO

ISO serves coordinating and distributing functions which the paying members of the organization manage — somewhat like the United Nations. ISO has 'civil servants' whose salaries result from member dues. Examination of the budget indicates the sources of income and the support given to various programs. Currently, 50% of ISO's income is from selling information technology standards. But there is no clear reason why the selling of these standards should be in the best interests of the members of ISO. Standards promoting organizations, such as ISO, facilitate the development of standards, but volunteers do the bulk of the work and no royalties are paid to those who have developed the standards.

One method of economically distributing standards uses the Internet. Costs associated with providing a 'standard documents server', its availability, and its maintenance will replace the paper-related costs. Should the dues of the members cover these functions? ISO could make standards freely, electronically available.

Consortia

Consortia may form in response to intense competition among the largest companies, each trying to dominate in volatile market segments, such as Unix platform sales. Despite their overtly commercial purposes, these organizations frequently find the development of de facto standards crucial to achieving their ends. In some cases, de jure standards bodies accept de facto standards or link them to their own standards (Updegrove, 1993). Should a standards organization take the product of a standards consortium and sell that product for a price?

To fund their development activities, consortia may charge fees to members. Funding levels of some consortia is on the order of millions of dollars. These funding needs can be either met by enrolling many members or by requiring very large contributions from individual members. For example, sponsor-level membership in PowerOpen (whose main goal is to foster rapid porting of software to the PowerPC environment) requires $250,000 in annual dues. Many consortia charge different rates for companies with different revenue levels, to permit smaller companies to participate. Again there is no relation between these costs to produce a consortia specification, and the cost of the eventual de jure standards. The consortium wants to obtain support for the standard, not to make money from the sales of the standards document.

The Government's Role

Recently a group of vendors and users in the United States made a request for free or sharply reduced prices for standard documents and asked the US Federal government to look into this need. Hearings conducted by the US Congress noted the possibility of competition among the US standardization promoting organizations for the returns from standards sales. Did such competition lead to reduced rather than increased cooperation among them? Did the high prices of the standards necessary to support the overhead of these standardization groups inhibit their use? A similar examination of this connection between standards and their prices could be made in Europe. (For mandatory standards the question of usage being inhibited by cost is not a relevant question. Thus, for Japan, which emphasizes mandatory standards, compliance and standards prices would have impacts different from the USA which tends to have voluntary standards compliance).

ISO, IEEE, and some other standards organizations are using the various copyright laws enforced by government to exploit a monopoly on the sale of some standards which, in some cases, were developed by volunteers who paid their own participating expenses. Monopolist or cartel practices can lead to good results, only if the monopolists are regulated. In democracies this regulation must come from the government. Unregulated monopolies or cartels have no need to consider "right-sizing," overhead costs eating into profit margins, or whether technological advances eliminate the need for continuing certain practices and staffs. Organizations such as ISO are essentially monopolies or cartels whose regulation is essentially internal as it comes in the form of the guidance from the members of the organization who pay dues to help support the organization and determine its policy. Why should organizations such as ISO be allowed to sell standards that were developed for wide, public dissemination by volunteers?

If ISO were to change its policy and if information technology standards were made freely available on the Internet, what would happen to ISO? The members of ISO would have to provide further direct membership funding to ISO.

The United Nations does not charge for its services to those who receive them. Rather the members of the UN have decided to pay dues so that services can be offered on a worldwide basis as needed. Why should ISO be different? The national members of ISO could pay enough dues to cover the administrative costs of ISO. ISO standards could be made freely available on the Internet. Paper copies could be purchased at a price enough to defray only the direct costs of providing the paper copy.

The analogy to the traditional publishing industry may suggest another model, namely the current model of strict copyright rules and priced products. The publishing advantage is the clear sense of feedback as to which products are wanted by the public and the support of a publishing organization that provides a kind of quality control and provides long-term inventory, marketing, and distribution. The standardization groups have, however, different objectives from the private publishing business, as national and international standards documents are more like government documents than documents produced by publishing houses. Among other things, the actual authors of standards do not get royalties.

CONCLUSION AND FUTURE

Information technology standards will continue to change quickly. The traditional standards development organizations' methods of working were too antiquated for information technology standards. These organizations are now changing to deal with the new needs. However, more change is needed.

Speed, the Internet, and Free Standards

The Internet is an excellent example of a technology that brings a broader policy issue to the fore. Namely, the issue of free distribution of standards is made particularly prominent by the availability of Internet distribution channels. Furthermore, the Internet itself exemplifies how voluntary, consensus-developed standards can succeed.

Information technology standards or, at least, their intermediate forms should be freely available on the Internet. Member dues for standards organizations should be high enough so that such increased accessibility of standards information could be provided on the network. This would be one way to increase the accessibility and utility of these documents and to create in the long run a more responsive and meaningful library of standards documents.

At the moment standards organizations often insist on copyright on documents which are drafts in the path towards a standard and will not release them to the public. Fortunately, some standards organizations are realizing that these documents should be available to the public. However, these same organizations may be expecting people to pay for access to these drafts. In the long run, such a mechanism for earning money from the standards may do less to support the long run of standards development than to allow free access across the Internet to these documents. The wide distribution of intermediary forms might stimulate interest in and compliance with standards.

Shortcuts and Consensus

In the discussion about Sun's shortcut for Java, everyone has agreed that Java is important and should be standardized. If Sun is successful in getting Java approved as an international standard, then Microsoft and many other companies might follow with their own submissions to make international standards of their products. Would this be good for interoperability in the information technology world or not? If Sun is unsuccessful in transposing Java into an international standard and continues to push its commercial success without regard to the international standards process, what will be the long run impact on the diffusion of information technology?

Should a private company be allowed to be a PAS Submitter? The community agrees that the traditional JTC1 processes are often too slow and too disconnected from marketplace concerns. Working quickly, however, conflicts with allowing all possible interested parties to have multiple opportunities to discuss changes. When a single company is a PAS Submitter, consensus building has been reduced in favor of private marketplace concerns and speed.

What alternatives are open to JTC1? Organizations can now apply for permission to operate as a new kind of standards developer by being PAS Submitters. In a sense, JTC1 awards a limited franchise. JTC1 profits from the sale of the standards that the franchisee makes, but JTC1 does not have to manage the day-to-day operations of the franchisee. JTC1 oversees quality.

Different organizations form to develop standards that might be later submitted to JTC1. One organization (IrDA) consists of over one hundred companies that deal with infrared devices; another organization (IETF) has now two thousand individuals attending its meetings and each wants to have some say about internet standards; another standards organization (IEEE Standards) is part of a professional society. The variety of organizations advancing standards is large and their way of working varies substantially. Should the right way to develop standards for one community on one topic be the same as for a different community and a different topic? Perhaps JTC1 should focus on a broker role in which an organization that wants to advocate a standard is connected with an organization that is well prepared to help develop such a standard. JTC1 would still monitor quality.

Consortia, such as PowerOpen, X/Open, and OSF, are not in a position to produce de jure standards. What difference does it make whether the standards are de jure or de facto? In the international community, all nations that agreed to the General Agreement on Trade and Tariffs (GATT) recognized that standards could be used as a non-tariff trade barrier and included remedies in GATT to provide protection against this happening. GATT calls for all nations to use ISO standards as a first choice and, for exceptional and documented reasons, use national or regional standards as an alternative. Standards organizations earn their right to define standards by using procedures that ensure the broadest consideration and widest acceptance of any resultant standard. Therefore, while smaller, more efficient organizations can develop de facto standards, the step of turning them into de jure standards remains the role of standards organizations like ISO.

The major standards development organizations, such as ISO and CEN, are making efforts to work with other organizations that have sound technical input, keen attention to market need, and operate quickly. ISO and CEN should continue in this direction but be careful to continue to be a filter for quality and consensus. Furthermore, standards development organizations such as ISO and CEN receive substantial funding from governments and should be able to provide standards documents for free on the Internet as a way to increase the impact of the standards that they sponsor.

REFERENCES

Dictionary of Units, (1999) 4th Edition, prepared by Center for Innovation in Mathematical Teaching at University of Exeter, Exeter: England, retrieved March 1999 from http://www.ex.ac.uk/cimt/dictunit/dictunit.htm.

European Committee for Standardization (1999), Brussels: Belgium, retrieved March 1999 from http://www.cenorm.be/.

Internet Engineering Task Force (1999), Reston: Virginia, retrieved March 1999 from http://www.ietf.cnri.reston.va.us/home.html.

ISO (1999) *Stages of the development of International Standards*, Geneva: Switzerland, retrieved March 1999 from http://www.iso.ch/infoe/proc.html.

ISO/IEC Directives (1995) *Procedures for the technical work of ISO/IEC JTC 1 on*

Information Technology, Supplement 1: The Transposition of Publicly Available Specifications into International Standards, Third Edition, retrieved March 1999 from http://www.iso.ch/dire/jtc1/suppl.html.

Rada, Roy, Carson, George S, and Haynes, Chris (1994) "Standards: the Role of Consensus", *Communications of the ACM, 37, (3),* 15-16.

Rada, Roy (1995a) "Sharing Standards: Consensus versus Speed?", *Communications of the ACM, 38, (10),* 21-23.

Rada, Roy and Berg, John (1995b) "Standards: Free or Sold?", *Communications of the ACM, 38, (2)*, 23-27.

Rada, Roy "Corporate Shortcut to Standardization", *Communications of the ACM, 41,(1)*, 11-15.

Smith, Mike (1999) "You know ISO... but what are PAS, TS and ITA?", Geneva: Switzerland, retrieved March 1999 from http://www.iso.ch/presse/pasetc.htm.

"Standardization Activities" (1994) *Computer Standards & Interfaces, 16*, 375-383.

Subcommittee 24 of JTC1 (1999) London: England, retrieved March 1999 from http://www.bsi.org.uk/sc24/index.htm.

Updegrove, Andrew (1993) "Forming, Funding, and Operating Standard-Setting Consortia" *IEEE Micro*, 52-61.

World Wide Web Consortium, Cambridge: Massachusetts, retrieved March 1999 from http://www.w3.org/.

Acknowledgements

The information for this chapter was largely based on columns that this author has co-authored for the *Communications of the ACM*. Those columns are cited in the references, and thanks are offered to the co-authors for their help in the past.

Chapter III

The Standardization Process in IT — Too Slow or Too Fast?

Petri Mähönen
VTT Electronics and University of Oulu

INTRODUCTION

Information Technology is one of the most rapidly developing areas in the technological world. In fact, the classical information technology that has been based on computer technology is now very quickly converging with telecommunication. Wireless networking, especially mobile telecommunications, has been one of the leading areas in business and technology during the last five years. The state-of-the art in mobile telecommunication has reached a point, where new products are introduced within a cycle of one year. The next big step shall be the introduction of wireless multimedia services through next generation wireless networks. The standardization of wireless multimedia systems is a very difficult task, as the standards shall include the crucial points in the two domains, telecommunications and information technology. It is becoming more and more difficult to recruit enough expertise for the technical committees to produce high quality standards in the converging information technology and telecommunication market domain. In this chapter, we will be using this domain as an example for the standardization difficulties of future.

The reasons have to be strong for initiating a standardization process, especially concerning industry, due to the fact that it is rather expensive to produce new and extensive standards. It has been estimated that the expenses of developing a single part of the Ethernet standards amount to approx. $10,000,000 (Spring & Weiss, 1994). The main development costs arise from the time, travel and salaries of the committee members. The standardization expenses are usually prohibitive for small and medium sized enterprises (SMEs). If the standardization organizations and committees can not guarantee that the investments done towards standardization are worthwhile we will have lots of problems in the future.

The organization of this article is the following. First, the main problems identified in the IT standardization process by several authors (Oksala, Rutkowski, Spring & O'Donnel, 1996 and references therein) will be discussed. Next, it will be explained why the standardization process is, nevertheless, important, and how attempts have been made to use the process for "standardization wars". The article will be concluded with a number of suggestions, which might help to improve the standardization process in the IT sector.

BABEL OF STANDARDIZATION

It is often very hard for people outside the standardization committees to understand the standardization process. The process can be highly complex (Cargill, 1989; ISO, 1992) and hence it is often seen as an overly political and bureaucratic issue. The nomenclature of standardization has been referred to by an author as "monk Latin and eastern mystique that can not be penetrated by industry people." Below, you will find a list of the main complaints against standardization.

1. The difficulty to understand the decision making process and the special jargon typical of standardization. This is one of the most often stated complaints and it can be overcome only by improving the communication. Standardization authorities should pay more attention to public relations on their list of priorities.

2. It has also been argued that standard texts are too complicated to understand. Part of the claims that standards are too difficult to implement stem from the difficulty of understanding standard texts—not from real technological problems within standards. SMEs, in particular, are concerned about this matter. Larger companies and academic institutes seem to be more tolerant towards large and complex standard texts.

3. Standardization committees have been claimed to produce too little and too late. It is widely believed that especially in the Internet era, the standardization process is far too slow to produce anything relevant for business.

4. Academic institutions and technological companies are also complaining that the consensus making process leads to less than optimal solutions. In some specific cases, it has been claimed that standardization can even destroy potentially valuable technologies through the tardiness of the consensus process.

5. IT-industry sometimes considers standardization as a danger to the IPRs (Intellectual Property Rights). For example, Microsoft has seen the Windows Application Programming Interface standardization by ECMA (ECMA - European Association for Standardising Information and Communication Systems; http://www.ecma.ch/) as a possible infringement of their IPR. Hence, Microsoft has threatened to sue ECMA and has also warned ISO of a possible infringement in the case of WAPI standards (ECMA 234) (Cargill 1997).

Let us comment each of the arguments mentioned above. There is no doubt that the standardization process is complex. The jargon inside the organizations is sometimes complex, although it can be claimed that it is not any more complex than any professional society jargon in the fields of engineering and science[1]. A good example is the use of the full official designation ISO/JTC1/SC22/WG5 instead of the term "Fortran technical committee". However, to make issues understandable for wider audience, we should not forget that simple language is often called for in public presentations. The lesson for standardization committees is that public awareness of the standardization should be emphasized more.

The complexity arises partly from the sheer number of standards. It is estimated that alone in USA there is a total of 93,000 standards in an active status (Toth 1996). Approximately 49,000 of these standards are non-governmental standards. Toth is also quoting that there are as many as 700 standard developing organizations in USA. These numbers show that standardization is a big business. It is no wonder that for several companies and people it is almost impossible to be aware of what is standardised and where. At VTT, we have estimated that when developing a *single* multimedia capable terminal equipment the number of standards one needs to consult is likely to be between 20 and 120. The expertise needed for following these standards and knowing what standards are relevant is becoming very expensive.

The role of different standards organizations are also commonly misunderstood. The official governmental or international bodies such as ISO (International Standardisation Organisation) and ANSI (American National Standards Institute) are sometimes confused with the activities of professional organizations (such as IEEE[2]) or purely industry based associations, such as DAVIC (Digital Audio Visual Council, www.davic.org). Especially in the field of telecommunications, standardization used to be the province of international organizations such as ITU (International Telecommunications Union), ISO and IEC (International Electrotechnical Committee). Now, the activities in telecommunications, information technology and multimedia are also addressed by a multitude of other players in the field. The standardization organizations can now be categorized into two main groups; formal (*de jure*[3]) and informal consortia (*de facto, grey or ad hoc groups*). The formal standardization processes are handled by traditional standards development organizations (ISO, ITU etc.), scientific or professional societies, trade associations or industrial standard organizations that can have a liaison with formal official bodies. Informal standards, in contrast, are produced by market forces (*de facto*) or by specific groups or consortia working independently (Toth, 1996).

It is a fact of life that technical IT standards tend to be complex. However, there is not much we can do to alleviate this problem. A good standard should not be open for too wide interpretations or misjudgments. Hence it is often impossible to write short and "easy" standard texts in the field of information technology. There is a strong case for preparing as appropriate standards as possible and, at the same time, for making the standards "as readable as prose". However, the strive for easier comprehension, although improving the legibility of the texts, might risk the quality of the technical standard, thus making the 'popularization' of texts a very dangerous road to take for standards committees.

The complexity arising from the number of standards and standardization bodies is usually referred to as *the Babel of standards* by the author. As mentioned before, the number of standards needed for building e.g. multimedia terminals can be very large. The table in Figure 1 shows some of the quite large number of different standards relevant to multimedia terminals produced between 1990-1995. Although the list is not complete, it gives a good example of the expansion of standard market.

Perhaps the most common argument against IT standardization is that the

Raster Graphics	▲▲▲	○▲		□	○□●	○■■	▲○■	■○
Vector graphics	■▲▲	▲			●□	■■		
Graphics metafiles	▲▲▲ ▲			■	▼▼			
Digital video	▼	○▼		▼	▼▼▼			
Analog video		○○□			▼	▼○○	▼▼▼	
Video audio mix	▲▲▲				■■■ ■○	○○○ ○○○ ○○○ □▲▲	■■○ ○○○ ○○○	○○○ ○○
Digital audio	■●	○○	○				○○▲	
Multimedia scripting	▲		■	■▲	▲▲			■●
Text and hypertext	■■■ ▲		■●	■□	■	○●● ▲	■■▲ ▲	
Optical media	■■■	■●	■●●	■■■ ●●	▲■■	■■		□
Distributed multimedia environment	*○○ ■●● ●	■■■ ▲	▲■□ ●	▲▲▲ *	■■■ ○●	*■■ ■■■ ■●▲	▲▲▲ ▲■●	■▲

■	○	●	*	▲	□	▼
ISO	ITU	ANSI	IEEE	Industry	DoD etc	Others

Figure 1: Illustration of the expanding number of relevant standards for a single multimedia product design. The list is not complete and refers to only one product example.

process and the time to achieve consensus is so long that the technology will have become almost antiquated by the time the standard is published. There is some justification for this line of argumentation. Because of this criticism, several standardization organizations have been re-engineering their processes. However, it is imperative that a certain level of maturity is achieved before any technology is standardised. It could be argued that if something is moving so rapidly that it can not be standardised, it is not worth being standardised anyway. Rather than being a prize to win, standardization has to serve some real purpose.

The consensus making process is, in fact, a very important test of maturity. If consensus can not be reached easily, it is quite clear that the proposed technology is not mature enough or other business-related issues are dominating discussions so strongly that standardization is impossible. It is also important to understand that the standardization is not about seeking the best solutions or maximally optimal technologies. This is a mindset that is very understandable for people with academic or engineering background. For many people, it is a hard lesson to learn that standardization is very closely related to the commercial decision-making done in industrial companies with products. A good standard does not necessarily lead you to the best possible solution, but instead the solution should be as good as possible, while allowing consensus, i.e. balanced view between differing opinions, to be achieved without destroying the whole purpose of the technology. As soon as a new committee member understands this, he will begin to enjoy the standardization process more than before. While it is essential to understand this aim, there will inevitably be some politics involved in the standardization work.

The standardization work even at the technical committee level is not related to R&D work or innovations. Innovations are submitted to the standards process, wherein we are "freezing" the state of the art to standards. Hence, the criticism from strongly research-oriented people is somewhat unjustified. Standards might include the best state of the art available, but they should not drive towards new innovations within the standardization process itself. For example, in the case of Internet standards or especially with MPEG-multimedia standards, there is no doubt that many standards include innovations (some of them brought about during the standardization, while the initial spark for R&D might have partially become from standardization). In conclusion, the place for innovations is outside the formal standards meeting — standards just agree on the solutions.

Standardization by Industrial and Special Interest Groups

There is a clear tendency towards standardization carried out by different industry groups in the field of information technology. During the past few years, a great number of industry lead standards organizations or "Forums" have been formed. A few examples of the kind:

- ADSL Forum that is focusing its standardization efforts on asynchronous digital subscriber line technology for fast copper-line connections (www.adsl.org).
- DAVIC (Digital Audio Visual Council) that is involved in the distributed multimedia transmission over different media, such as cable and wireless connections (www.adsl.org).
- ATM Forum that is doing major work in the standardization of asynchronous transmission technology.
- BlueTooth is a completely industry-lead consortium focusing on the low-power short-range wireless communication standard for connecting devices, such as portable computers and cellular phones. It was formed by Ericsson, Nokia, Toshiba and Intel, now having has more than 200 members (www.bluetooth.com). Strictly speaking, this special interest group is not even a standardization forum, as only the founding members have a vote and form consensus between themselves.

The standardization work done outside the more formal international organizations varies in quality. It should also be borne in mind that the aims of these groups are likely to be very different. The common belief is that the work can be done faster by the special interest groups than by a formal standardization organization. Some groups are aiming at producing standard texts that are then submitted to a national or international standardization organization for approval. On the other hand, some other groups do not care about the official standardization. They are producing *de facto* or *ad hoc* standards that might be voluntarily adopted by industry or other standard bodies. Usually the variation in the quality of work is greatest in the latter groups. Finally, some of the forums are not "democratic", the voting power possibly being limited to certain founding or paying members. Quite often industrial special interest groups lack academic members. Paradoxically, these industry-only standardization committees are not always more effective than their official counterparts with open processes and academic members.

Internet Engineering Task Force (IETF) and related organizations are among the best performing "hacker" or grey standard communities in the field of standardization. The process is very open, discussions and opinions are effectively exchanged via Internet, and everybody can contribute. The final decision-making is less open, but still following democratic principles and, again, highly effective. In the case of information technology standardization, we can learn a lot from the IETF approach.

International Telecommunications Union (ITU) and European Telecommunications Standardization Institute (ETSI) are among the major players that are standardising telecommunication technologies. They are formal bodies, in which big telecommunication companies, in particular, are trying to form consensus to build up interoperable next generation telecommunication systems. There quality of work done in ITU and ETSI is very high and they follow strict procedures in preparing the final standards. Although they are slow, it is for good reason. One has to bear in mind that when a standard is finalised and agreed upon, telecommunication and IT-companies are expected to make major investments on the final implementation and deployment of the standardised systems. Hence, one would like to be as sure as possible on the fact that the standardised architecture and system are as mature as possible before committing to multi-billion dollar investments.

Standardization Wars: Case study of Java

Sun Microsystems applied for the status of Publicly Available Specification (PAS) submitter from ISO to standardize Java. The PAS process was introduced by ISO/IEC JTC 1 as a part of its process streamlining. The justification for PAS was to introduce a mechanism to for adopting de facto standards that having the same open characteristics as the usual IT- standards. The main purpose is was to speed up the standard process for "almost ready" standard proposals. This application was done made in March 24-28, 1997. The move was somewhat unheard of, because the PAS status was thought to refer to impartial standardization groups with due processes outside ISO/IEC JTC1. Traditionally, PAS submitters were groups such as ECMA and X/Open. Sun was claiming that it had impartial open processes already in place and referred to the need tool for standardizing the Java technology as fast as possible. The discussions and negotiations, including two full country ballots for PAS status, took time and it was not until November 24-28, 1997 that SUN was (conditionally) approved as a PAS submitter (for details see e.g. O'Gara, 1997). The conditions stated by several countries referred to openness, standard maintenance and trademark issues that had not been resolved during the ballot. When writing this chapter (1999), Java had not been submitted for an ISO process and the standardization has not been completed. Hence, the justification for allowing a single for-profit company to use the PAS process, because of reasons of claimed speed-up, is still an unproven concept. Within two years, the formal ISO/ IEC JTC1 working group would most probably have produced a draft standard for approval. The latest developments in the Java standardization case have been unfolding during the proofreading phase of this chapter (June 1999). Sun has claimed that ISO/IEC JTC 1 has changed rules of PAS submission during the submittal process. Now, JavaSoft (part of Sun) will not submit anything to JTC 1, but will, instead, use ECMA as their standardization route. The comments within the standardization community have been very fairly mixed. Some members have been very frustrated and there have been questions whether "Sun is playing games... and is it really interested in standards at all?", or "do these Java people at Sun understand the standardization process and organizations at all?". Some people have shown understanding for the difficulties of JavaSoft; others have felt that JavaSoft has been "using" them in a public relations and marketing ploy.

In any case, the fierce discussion around the Java PAS submissions has shown that standardization sometimes amounts to an outright war between giant companies—and sometimes even between companies and standards authorities. In the case of Java, especially Microsoft and Sun Microsystems have been pushing their point of view rather forcefully onto national standardization bodies. There is no single "good guy" in this particular case. In fact, the whole issue has been a bit discomforting for the long-term standardization committee members. Both sides in the argumentation have been using lots of "legalese", i.e. submitting fuzzy or marketing-like texts and arguments to support their opinion. In fact, the arguments of companies concerning the politics involved in the work of international standardization committees tend to sound hollow after what the companies themselves have exhibited, e.g. in the case of Java and ECMA-WAPI standardization.

The main issue in the acknowledgment of the PAS status for a for-profit company is openness. The PAS process by a single vendor does not fully guarantee an open consensus process. Sun has been reporting in its comments ("The Responses" 1997): "We systematically distribute the first draft of any specification to all our licensees" and "When we placed the Beans specification on our web site...[it] was accessed 40,000 times...and we received numerous email comments, all of which we read" (James Mitchell comments see "The Responses" 1997). The problem in this answer is that a due (consensus making) process is not contained in "we distributed it to *licensees*" or "which we read". In this form of process there is no guarantee for any discussion on the comments, or there is no real aim to make a consensus out of the comments. This is the central question, although also the questions on intellectual property rights (such as Java trademark) are very important during the country ballot—and these still remain unsolved.

However, although the Sun position seems unjustifiable, it should not be forgotten that a due process at the ordinary technical committee level, e.g. in ISO or ITU, is not completely democratic either. One can submit proposals or comments to working groups, but without a champion within the group there is only a very slight possibility for them to be passed for the standard. To be objective, due to the fact that the participation in the working group is rather expensive, it could be admitted that the only difference is to be found in the level of openness. Thus there is a possibility that a compromise over the Java issue can finally be found. In conclusion, the main lessons learned from standardization wars are; (a) standards are important, (b) even PAS or industrial processes are not necessarily faster than committee work, and (c) industry comments on overwhelming bureaucracy, legalese and politics in committee level can be ignored as industry is behaving just as badly as any standards committee or authority.

Fighting over the standardization issues is a common phenomenon, especially in telecommunications, where the standard decisions may affect billion dollar businesses. Thus the power politics involved in the standardization processes within IT and telecommunications domains should be seen as an inevitable part of the process. Any attempts to get rid of it would be considered naïve and unrealistic.

THE BENEFITS OF STANDARDIZATION

There are ultimately several different approaches in IT standardization:

1. To produce absolutely necessary *de jure* standards. Good examples are standards for telecommunications systems, such as GSM, UMTS and IS-95, or programming language standards. In these examples the standardization is absolutely necessary, as without it there would be no global business or user-activity. For example, in the case of mobile communication, the firm standard is needed for ensuring interoperability, frequency licensing etc., before large investments are made. Quite often intellectual property rights are playing an important part in the development of standards, as patent free technologies should be encouraged, but many companies do not like to give up potentially valuable intellectual property to common use. However, quite often compromises over IPR policies can be found. In case of prior IPRs, owners are offering licenses to other standard users for a "fair sum," on a non-discriminatory basis and/or on "reasonable" terms and conditions.

2. Useful *formal* standards are justified by the fact that the activity or business will be stronger, if there is a strict standard available. Most of the programming languages actually belong to this category. It is good to have interoperability between compilers and a good standard will foster the interest for a wide use of the language . However, as in the case of Java, the *de jure* standard is not really needed for making a language widely accepted. A strong standard might, even in the latter case, prove an important tool for *protecting the cleanness of the various implementations* (i.e. concerning compatibility between different vendors).

Necessary *de facto* standards, such as TCP/IP for Internet. It is a well known fact that TCP/IP quickly gained dominance over the formal OSI layer architecture. Networking standards are necessary, but, as it can be seen, a *de facto* standard can be adopted, because it works, it is totally open and after market adoption it can prove stronger than a weak formal standard.

Naturally, there are also other approaches, but these three examples do show the essential fundamental differences. Usually there is competition between different standards, and finally, it is the market and end-users that make the choice among the standards. The main special feature of telecommunications, and sometimes of multimedia systems overall, is that often a standard is necessary for the opening of business markets. Hence, it is evident that standardization in the field of information technology is absolutely necessary.

It is also essential to be aware of the different levels that exist within standardization. Standardization can be used for many purposes. If a standard defines a specific *physical* aspect of a device, such as the size, mechanical dimensions of connectors etc., we call it "physical standardization." Although physical standardization used for providing *sameness (uniformity)* of products might be important in many cases, it can not be regarded as a universal requirement. In information technology and telecommunications, as well as in multimedia, the critical issue is not to promote sameness, but to provide *compatibility* (Krechmer 1996). For example, in the MPEG video compression standard, the idea is not to define an exact algorithm for the industry to build video encoders. It is up to the industry and competition between implementers to work out the best or the most suitable implementation. A similar approach is taken with e.g. programming languages, where the standards committee is very careful not to give any implementation level guidance. This is the fact that makes information technology standards differ from physical device standards.

RECOMMENDATIONS AND DISCUSSION

Oksala et al. (1996) have pointed out that the composition of committees is an important question to resolve when one wants to know how they are working. The distinction between the technical work and the approval process is also a very important one. One of the main arguments against some IT standardization work, especially from academia, has been that the technical committee members do not have enough to say in the approval phase. However, to keep a process open, people outside the technical committees should have an opportunity of voicing their opinion. It can be argued that although the technical committee level should be composed of mainly technical experts, in most cases it can be recommended for the approval process to be carried out at the national level by people who have not been actively participating in the technical committee's work. This will make the approval procedure more impartial, and it will guarantee that no single powerful company can begin to dominate the standardization process at the committee level. At the moment the parties with the widest representation in IT standardization are SME companies and universities.

One of the key standard processes in digital multimedia has been that of MPEG2. In MPEG2, several areas have been left *intentionally* unspecified. These unspecified details can be found in the areas of conditional access, transport layer features and stream filtering requirements. The decision to leave these areas open made the standardization process considerably easier Hence, a lot of room was left for implementation issues in the standard. In future we should be prepared to make quick standards which involve leaving some issues open. It is better to have something agreed upon and ready for implementations than nothing at all.

There is a strong case for setting up a central database organization for IT and telecommunications standards. Although complexity is an inherent feature of standardization, I think that it would be very beneficial if we could find or form a nonprofit organization for developing and maintaining a web-based service including a database of the main standards and committees involved. This database could be used for finding out relevant standards for specific applications, including cross-references between standards. The database effort could be financed by large international organizations and its use should be free of charge for most end-users.

The standards should be made available on the Internet. There should be a possibility of gaining free access to at least part of the texts, to be used in teaching and research, especially in third world countries. Although the financing of the standardization work is very important and the present prices reasonable (although not cheap), these prices can be prohibitively high for educational institutions in developing countries. IETF (Internet Engineering Task Force) is a good example of this front. All standard proposals and main standards are available at their web-site (www.ietf.org).

The good step towards a better understanding is GII Virtual Roundtable (www.globalcollaboration.org). This is an organization-neutral online forum that includes a large set of standardization organization such as IEC, IrDA, ISO, JTC 1, POSI, VESA, ETSI, DAVIC etc (Ryan, 1998). The goal is to provide a common forum for users and consumers alike for voicing their opinions. However, this is just the first step, while collaboration and open electronic archiving should be extended and encouraged further.

The standardization authorities and organizations should be more proactive toward emerging technologies and standards consortia. Their work programs should involve more cooperation and prevention of overlapping. The competition between different committees and organizations presents, in my opinion, one of the greatest dangers for international standardization. In 1987, ISO and IEC set up a Joint Technical Committee 1 (JTC 1) by integrating the technical committees of the ISO and IEC. The purpose of this cooperation was to eliminate the serious overlap between ISO and IEC in the information standardization. The ISO/IEC JTC1 now includes 18 subcommittees. This sort of streamlining should be strongly encouraged.

Unnecessary grey and ad hoc standardization should be discouraged. Formal standardization organizations with strong market-driven approaches should be encouraged. Professional standards authorities give are likely to yield better long-term value, because (a) they integrate their activities with the work of other standards authorities, (b) have consistent rules of dealing with intellectual property issues, and (c) a they employ a set of consistent rules are employed for maximising fair and unbiased operations to produce good quality standards with maximal consensus (Krechmer 1996).

The JTC1/SC22 committee (for coding of audio, picture, multimedia and hypermedia information in JTC1; one responsible for MPEG-standards) has introduced a call for proposals phase for its process. This method was pioneered in the development of MPEG-1 and MPEG-2, and the people introducing it should be congratulated. The idea is that after the definition of requirements, an *open call for proposals* is issued asking for relevant technology for meeting the requirements. After the call is closed the proposals are evaluated, and the most promising technologies are chosen for further development. Several calls can be issued within the standard development process. This method has proven promising for getting more involvement and a better input to the standards process. It is clearly related to the Request-for-Comments (RFC's) process of IETF. The main difference is to be found in the fact that RFC can be submitted at any time and without any formal call to IETF, while the RFC editor will decide whether the submitted document will be included in the RFC base. Both the Call of Proposals and the RFC have a proven track record and merits, and they could be included in most standardization processes.

The engineering training should include at least some basic teaching concerning standardization. At present, it is extremely rare to see any courses at undergraduate or graduate levels that would give even rudimentary ideas about standardization. The standardization organizations, processes and the main standards should be included in the curriculum. This will not call for too much work, and some of it could even be embedded in already existing courses, as the author has been doing. In any case, the hopes of gaining a better understanding, better cooperation or more funds for standardization will be greatly encouraged, as soon as we start giving our students more information about these issues.

Finally, funds should be made available for people from SMEs and academic institutions (especially people from universities) to make it easier for them to participate in the standardization work. National standardization offices and other funding agencies should jointly set up a national funding authority, which would provide the under-represented participants with financial help so as to facilitate an active participation in the standardization processes.

CONCLUSIONS

As mentioned in the introduction, the IT standardization process is essential for securing not only the coherence of the R&D work worldwide, but also the compatibility between different vendors and service provides. There is no need for a complete unification of the different standardization approaches. Complexity and slowness are likely to be inherent features in the standardization process. However, public awareness, a central database with all standards and with a limited or full access to standardization texts for everyone through Internet should be very carefully considered.

To conclude, I will now try to answer the question made in the title of article. IT standardization can be too slow and too fast—depending on the situation. However, the benefits of quickness tend to be greatly overemphasised, while the slowness is, at least partially, a myth (in the sense that commercial companies or groups always do a faster job). Hence, IT standardization is useful, and the people participating in the process have no reason to be shamed or shy about their work and achievements. In fact, thinking about some purely industrial special interest groups (even forums with liaison with official standardization agencies), these are sometimes paradoxically both too fast and too slow. Too fast, because they produce too quickly, lower quality standards, which are not mature enough. This becomes evident through great and quick changes between versions, clear mistakes in the standards and great difficulties in the implementation (a part of the standard may be even be conflicting with some other part); and too slow for the reason that these groups fail to follow the changes in the market and technologies any faster than formal standardization bodies. Due to the essence of these groups, their being narrow special interest groups, it is sometimes very hard for them to keep up with developments (e.g. DAVIC was not quick enough to see and to follow the transition from ATM-dominated paradigm to TCP/IP dominated networking.)

The main driving force in the future will be the convergence of IT-technologies and telecommunications. This will present a major challenge for the standardization work, as we have to be able to collaborate over quite a large technology domain. Cooperation, not competition, will be important. Because of this, a more professional and businesslike approach should be adopted in the standardization. In the long run, the so-called grey standardization where ad hoc or industry-interest groups are creating "marketing standards" is not likely to provide the best solution.

ENDNOTES

1. The disease of acronymization (i.e. inventing new acronyms out of the first letters of words) and jargon is very common in all branches of engineering and marketing. This phenomena is not related to standardization, but seems to be common cultural practice within different specialized societies.
2. IEEE, the Institute of Electrical and Electronics Engineers, is a professional society that is also producing standards. IEEE is member of ANSI.
3. Even in the case of formal (de jure) standards, organizations like ITU and ISO stand to the voluntary application of standards. It is very rare that actual statutes, laws or governmental regulations force to use standards. Statutes or rules are more common in case of safety critical systems, rules governing governmental purchasing (such as military equipment or projects) etc.

REFERENCES

Besen, S.M. (1990). The European Telecommunications Standards Institute, A preliminary analysis. *Telecommunications policy*, December 1990, pp.521-530.

Cargill, C.F. (1989). *Information Technology Standardization. Theory, Process and Organizations*, Digital Press, Digital Equipment Corporation.

Cargill, C. (1997). Sun and Standardization Wars. *StandardView*, 5(4), 133 -135.

Gibson R. B. (1995). The Global Standards Process: A Balance of the Old and the New, Standards Policy for Information Infrastructure, Cambridge, MIT Press

ISO/IEC (1992). *ISO/IEC Directives-Part 1: Procedures for the technical work*. Second edition. Geneva. (available at www.iso.ch/dire/directives.html)

O'Gara, M. (1997). The Analyst's View: What the World Heard. *StandardView*, 5(4), 195 -197.

Krechmer, K. (1996). Standards Make the GIH Possible. *IEEE Comm. Magazine*, August 1996.

Krechmer, K. (1996). Technical Standards: Foundations for the Future. *ACM Standard View*, vol. 4 no. 1.

Oksala, S., Rutkowski A., Spring M. & O'Donnell J. (1996). The Structure of IT Standardization. *StandardView*, 4(1), 9 -22.

Ryan, H. J. F., (1998). ISO/IEC JTC 1 Directions in Multimedia and GII Standards. *IEEE Comm. Magazine*, September 1998, p. 108-114.

Spring, M.B. & Weiss M. B. (1994). Financing the standards development process. In B. Kahin (Ed.) *Standards Development and Information Infrastructure* J. F. Kennedy School of Government, Harvard Univ., Cambridge MA.

The Responses Begin, (1997) *StandardView* 5(4), 146-151; also the submissions for ISO/JTC1 in 1997, that the author was reading as the Finnish representative for ISO JTC1/SC22 Java Study Group.

Toth, B. (1996). Putting the U.S. Standardization System into Perspective. *StandardView*, 4(4), 169 -178.

Acknowledgements

I would like to thank all my colleagues in different standardization organizations for enjoyable collaboration and useful discussions. I am also grateful to Mr. Jyrki Huusko for helping me in the preparation of Figure 1.

Chapter IV

Institutional Dilemma in ICT Standardization: Coordinating the diffusion of technology?

T. M. Egyedi
Delft University of Technology

INTRODUCTION

The effect of Information and Communication Technologies (ICT) on our daily life needs no explanation. That standardization reduces diversity, facilitates interoperability and thus plays an important role in diffusing ICT uses is clear. Standardization matters to the ICT market. Less evident is the way in which standardization influences the kind of ICTs that become available to customers. In the past it was common to view committee standardization as a locus for collective learning and exchange of technological knowledge. New standards were expected to embody state-of-the art ideas on technology. But practitioners, i.e. ICT standards developers and implementers, as well as standardization watchers repeatedly voice disappointment about the technical content of new standards and the process of committee standardization. They criticize the formal standards bodies for furthering a politicized mode of standardization, and point to the greater use of standards that stem from other arena (e.g. consortia, user groups, and practitioner organizations). These other institutional settings of standardization are held to produce more applicable standards and standards of better quality technology-wise. Are they barking up the wrong tree? In this chapter I explore whether this is the case.

CONCEPTUAL FRAMEWORK

Several studies on standardization have examined the impact of standardization on ICT development (e.g. Hanseth, Monteiro, & Hatling, 1996, Schmidt & Werle, 1998). They use Social Shaping of Technology theories to do so (e.g. Bijker, 1990; Callon, 1986; Hughes, 1987). In these studies the standards setting environment is—implicitly or explicitly—treated as a setting of technology development.

Standardization is an endogenous factor in technology development (Egyedi, 1996). It is a means to coordinate technology development (Schmidt & Werle, 1998). However, it proves to be difficult to pinpoint the impact which standardization has (e.g. does it hamper flexibility in technology development; Hanseth et al. 1996). One reason is that its effect depends on the maturity of the technology concerned. In the classic approach to technology development three successive phases are distinguished (Cramer & Schot, 1990). The phase of (1) invention covers all activities from the generation of an idea to the development of the new process or product. If successfully marketed the invention becomes an (2) innovation. The last phase is that of (3) diffusion of the innovation in the market. This distinction has been refined and criticized for several reasons[1]. For the argument in this section, it suffices to distinguish between emergent and mature technologies. In the field of standardization the issue of technological maturity has been used to sketch the dilemma of *when* to standardize. Early standardization, practitioners say, forestalls diversity but precludes experience with the alternatives, while late standardization makes it more difficult to reach consensus.

To complicate matters, in standardization too different stages are being discerned (Cargill, 1989; Bonino & Spring, 1991; Mansell & Hawkins, 1992). For example, Mansell & Hawkins (p.45) discern the planning stage (i.e. determining standardization priorities), the negotiation stage (i.e. as in committee standardization) and the implementation stage (i.e. the way standards are implemented). Each stage may affect technology development in its own manner. For example, a much-voiced view is that if a standard is widely implemented, compatibility frees resources for innovative activity.

In order to narrow down the issue, only certain phases in standardization and technology development are of immediate relevance for the criticism addressed in this chapter. Usually the discussions about the pro's and con's of standardization start out from the assumption that standardization deals with emergent technologies, and that the negotiation stage is the most influential stage of standardization. (I will review these assumptions in the discussion.) The negotiations take place within an institutional context. The context of formal standardization is held to be accountable for the lack of innovative and applicable standards.

Rommetveit shows that moving a decision process from one arena to another with different structural features changes its outcome.[2] Analogously, several studies argue that the institutional context of standards committees is highly relevant to the outcome of the standards process (Genschel, 1993, p.26; Bonino & Spring, 1991, p.102). The organizational procedures embed ideas on how standardization should proceed, beliefs on what is important in the process of establishing standards and why it is important, assumptions about the standards environment, etc. These shared ideas, assumptions, values and beliefs are captured by the term *standardization ideology* (Egyedi, 1993). Once ideas are institutionalized, they acquire a taken-for-granted quality and are not easily dismissed or changed (e.g. March & Olsen, 1989, p.52). The institutionalized ideology of, for example, the formal standards bodies, thus fosters continuity in the standardization approach. It indicates the role, which the formal standards bodies aim to play, and clarifies the direction in which institutional provisions influence current standards work.

Standards procedures regulate the committee process. Rules are laid down on how the negotiation of insights and interests should proceed. The *raison d' être* of formal standardization is that the standards resulting from this negotiation process will in many cases be different from the way they would have been without the regulatory role of standards bodies. The force field of vested interests and technology-related standpoints is transformed. The standardization structure and procedures direct the transformation process. As the rules of a game affect its outcome, provisions for formal standardization affect standards.

In the next sections, I examine the standardization ideology of the formal international standards bodies, the role they aim for and the procedural adaptations that have taken place in reaction to pressure from European developments and 'competing' standards arenas. I compare formal standardization with other forms of standardization, and conclude by readdressing the criticism voiced by practitioners.

FORMAL STANDARDIZATION

The formal international standards structure consists of the International Telecommunication Union (ITU), the International Electrotechnical Commission (IEC) and the International Organisation for Standardization (ISO) which address standardization of telecommunications, electrotechnique and a domain, encompassing all other areas of international standardization, respectively. In the domain of ICT the three organizations cooperate. Since 1987 the ISO and the IEC cooperate in the ISO/IEC Joint Technical Committee 1 (JTC1). The internal structure of the three standards bodies is similar, although the names of organizational entities differ. Each has an overarching forum in which standards policy is decided upon. The standards work occurs in technical committees, subcommittees and working groups. In the ITU-T (ITU Telecommunication Standardization sector), 'questions' are allocated to 'study groups'. These questions are processed in two stages. The first stage consists of drawing up the draft Recommendation (i.e. the draft standard). If the draft is mature enough, the approval stage is invoked. In comparison, ISO and IEC have a many-staged process (see Table 1). For more details on organizational structures and procedures, I refer to Egyedi (1996).

Project stage	Associated document	Name
0 Preliminary stage	Preliminary Work Item	PWI
1 Proposal stage	New work item Proposal	NP
2 Preparatory stage	Working Draft(s)	WD
3 Committee stage	Committee Draft(s)	CD
4 Approval stage	Draft International Standard	DIS
5 Publication stage	International Standard	IS

Table 1: Project stages and associated documents (source: ISO/IEC Directives, Part 1, 1992, p.16).

Ideological Rationale

The procedures of the formal standards bodies express values, assumptions and beliefs. For example, they embed democratic values (e.g. ISO/IEC's Public Enquiry allows all non-participating interested parties to comment on draft standards) and reflect the desirability of a technical and politically neutral standards process (e.g. in the approval stage only the negative votes which are accompanied by technical arguments are counted). Together such features make up the tissue of the standards ideology. The most pervasive ones are listed below.[3]

1. The *consensus principle* is adhered to in the most essential stages of decision making, such as the preparatory and committee stage in the ISO/IEC and the approval process in the ITU-T.
2. They stand to the *voluntary application* of standards[4], which is why the ITU speaks of recommendations instead of standards.
3. There is concern for the *quality of standards*[5], an element relevant in all three standards bodies.
4. Participation is based on *national membership* (national standards bodies) and not on direct membership of interest parties and companies[6]. ITU's membership of national administrations is similarly nationally oriented.
5. ISO/IEC strive for a *broad constituency of national delegations*[7].
6. They adhere to *democratic* working methods by means of a "well-balanced influence of national members" in management bodies of international standardization organisations and an "open, democratic process of decision-making"[8]. Democratic ideals are also evident in the ITU procedures.
7. All three bodies strive for an *impartial, politically and financially independent* organisation and procedures[9].
8. All three bodies strive for widely used and thus in principle *international standards*[10].
9. ISO/IEC procedures are designed to promote *fair competition and fair trade*[11].
10. ISO/IEC strive for *openness in information*[12].
11. All three bodies strive for rational, *technical discussion.*[13]

In the past, the ITU ideology differed in some respects from that of the ISO and the IEC. Irmer (1992) speaks of the 'closed circles' of the Study Groups. The ITU largely excluded competitive service providers. It did not aim at fair competition and fair trade (9). There was no 'broad constituency of national delegations' (5) or openness in information (10). Changes have been underway. The ITU is moving towards more openness and fair competition.

Ideological rationale. The set of ideological features reflects a form of argumentation. The rationale is that international standards cannot be imposed. If they are to be voluntarily applied, potential implementers need to be convinced that the standards have wide support. The ITU rallies support by way of international consensus. ISO/IEC fosters support by taking care that all interest groups have the possibility to state their requirements and influence the standards process. The spectrum of interests is represented by broadly constituted national delegations, according to the rationale. Consensus decision-making ensures that the utmost will be done to accommodate minority standpoints. Democratic and impartial working procedures heighten the acceptability of the standards. The measures serve to maximize the likelihood that formal standards will be preferred to proprietary standards and will be widely implemented[14]. The rationale underlying the standards system is that a democratic approach is the most robust approach. The ideology could be called a democratic ideology.

Friction between ideology and praxis. In practice, certain ideological objectives pose a problem. In the decision fora of the ITU there is a "heavy layer of international politics" (Rutkowski, 1991, p.292; Raymond, 1990). The vulnerability of democratic and impartial procedures (7) is capitalised on to further political interests. At stake are institutionally sanctioned strategies. Conflicting interests between economic regions and between the developing and the industrial countries politicize the standards process. In the ISO and the IEC the 'broad constituency of national delegations' (5) is difficult to realize. The composition of technical committees is more homogeneous than the ideological feature indicates. National delegations most often consist of industrial interest groups.

Process or Outcome?

Extrapolating, is the role of formal standards bodies to regulate the process or to produce standards? Standardization theory provides few clues on the matter. Turning to institutional theory, March & Olsen (1989, p.49) address this question with regard to political institutions. They question the primacy of action and outcomes. The core task of political institutions is to confirm the legitimacy of decisions. Legitimacy is furthered by demonstrating that intelligent intentional choices are made, by securing that relevant people are involved and by an appropriate control structure (March & Olsen, pp.50-52).

The same elements are evident in the standardization ideology. They define the role of formal standards bodies as guardians of the process. Standardization procedures serve to legitimize the process.[15] They are designed to regulate the process in a way that most participants will benefit from the result. As such, they interpose a democratic layer between the field of forces and the standard. Concern for the way that standards agreements are reached is what justifies the elaborate institutional context.

The formal standards bodies are, however, inclined to stress 'outcome' rather than 'process' results because of outside pressure. To show for their effort, they present statistics about the amount of work items tackled, the number of standards developed, — if favorable — the speed of their development, the amount of pages which standards documents cover, etc. Process results are seldom highlighted. More in line with their role, one would expect them to advertise process successes such as, for example, heterogeneous standards groups, number of 'true consensus' standards versus the amount of 'minimal agreement' standards, spread of implementations, etc.

PRESSURE ON INTERNATIONAL STANDARDIZATION

Although the traditional ideology of formal international standardization has remained rather stable during the 1980s and 1990s, their structure and procedures have undergone a number of significant changes. These changes were carried through in reaction to pressure from inside and outside formal standardization.

Firstly, in the 1980s the overlap between standardization activities in the fields of information technology and telecommunications became increasingly evident. It necessitated close cooperation. Institutional *integration* was settled with the instalment of the ISO/IEC JTC1 and the JTC1 - ITU cooperation agreement.

Secondly, in 1985 the European Commission laid down its plans to achieve a common market by 1992. It invoked the support of the formal European standards bodies. These were to deliver the large volume of standards required to remove technical barriers to trade. In reaction to the threat of a standards-based *'Fortress Europe'* and the haste demanded by 1992 time-schedule, the international standards bodies devised means to quicken the pace of standards production.

Thirdly, European fortification in the 1980s elicited responses from other regions as well. These intensified their standards activities in the field of ICT, especially with regard to 'open systems development'. Regional and *interregional alliances* were forged (e.g. MAP/TOP user groups, Workshops on Open Systems, organisations for telecommunication standardization). The international formal bodies welcomed the regional alliances on open systems as these strengthened their own activities in this field. The *inter*regional alliances posed more of a threat. The threat was dealt with in different ways. After some initial alarm, the interregional group on telecommunications, for example, was enrolled as a feeder-platform.

Fourthly, in the late 1980s increasingly multi-party standards from other sources than the formal international standards bodies were being used — notably Internet standards. The formal bodies devised ways to formalize these *grey standards*.

In particular the pressure from Europe and from grey standards bodies forced the formal bodies to rethink their role in international standardization. Both forces drew attention to a specific problem. The 'Europe 1992' schedule centralized the problem of timely standardization, while the activities of grey standards bodies drew attention to the problem of implementing formal standards.

Eurocentrism

In the events leading up to 1992, the European Telecommunications Standards Institute (ETSI), which took over the standards work from Conférence Européenne des administrations des Postes et des Télécommunications (CEPT) in 1988, played a central role. This new standards body had procedures that differed from the two other European standards bodies (Comité Européen de Normalisation (CEN) & Comité Européen de Normalisation Electrotechnique (CENELEC)) and from the formal international standards bodies. The most radical differences were that

---

Box 1: Institutional Concerns in International Standardisation
- Integration of telecommunications (ITU) & information technology (ISO/IEC JTC1) (1980-1993)
- Threat of 'Fortress Europe' and European standards (1985-1992)
- Rise of interregional networks on standardisation (1985-1993)
- Increasing significance of grey standards (1987-....)

individual companies could become full ETSI members, that the voting procedure was less susceptible to minorities opposing standards approval, and that project teams could be installed to develop standards. The European Commission applauded its standards approach and exemplified ETSI in the Green Paper on Standardization (1990).

In view of the size of the European market, the threat of a standards-based 'Fortress Europe' and the pace demanded by the '1992' objective, fortification of international standardization was called for. Three general strategies were pursued. Firstly, the international standards bodies reviewed and reorganized their internal structure and procedures. Secondly, they intensified their cooperation (e.g. ITU/ISO/IEC, 1993). Thirdly, cooperation with European bodies was established as a means to subsume European standardization. The measures showed a strong orientation towards Europe.

- *ETSI's membership rules and working methods were contemplated as possible alternatives.* Some ideas were adopted, such as the use of project teams (IEC, 1992). A proposal for non-national membership in the ITU (e.g. Irmer, 1992) was turned down by the Plenipotentiary.
- *The international bodies went out of their way to accommodate ETSI in their policy documents* (e.g. in the ISO/IEC Code for Good Practice for Standardization)[16].
- *European Directives influenced the international agenda and time-schedule.*[17]
- *The conditional nature of European cooperation with international standards bodies was accepted in the cooperation agreements concerning common planning of new work and parallel voting (1989/1991).* The agreements served as models for future agreements with other regions. (IEC, 1992, pp.13-14)

The European interests were well cared for. A form of Eurocentrism was at stake, driven by (see Figure 1):
- the Commission's Directives and mandates, which affected European and international time-schedules and priorities in standardization (3,6);
- ETSI and the exemplary function of its procedures. Its influence extended to other sectors within Europe (4), to the international standards arena (6) and —via interregional activity—to international telecommunications (5); and
- CEN and CENELEC's cooperation agreements with their international counterparts, which took on an exemplary status (7).

Figure 1 schematizes the line of argument. The Commission strongly influenced European standardization (3). The figure draws attention to (1) the enabling nature of the Directorate-General (DG) structure of the European Commission in the establishment of ETSI[18] and (2) the role of the CEPT therein. The Commission's requirements and ETSI's example affected both CEN and CENELEC's workings (4), and those of the international standards bodies (5,6). Thus, the European actors to a large extent shaped developments in the international standards network up to 1993 by communicating a sense the urgency[19].

Grey Standardization
A second influential source of change in the international standardization is

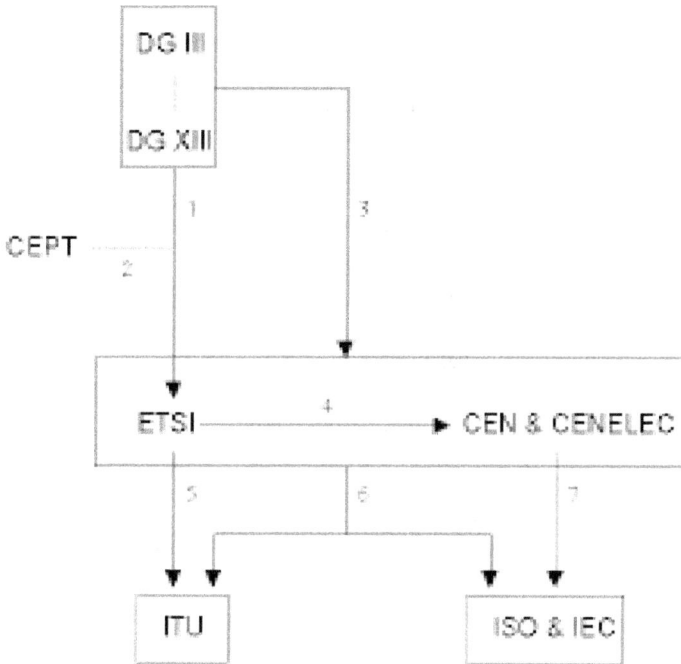

Figure 1: Eurocentrism in international standardisation.

grey standardization, a term first used by Bruins (1993). It became a force to be reckoned with roughly from 1987 onwards. It is currently still of high relevance. Grey standardization is typically initiated and driven by implementers. The development and implementation of standards occur in parallel. Standardization takes place in multi-party environments and leads to Publicly Available Specifications (PAS).

Grey standards bodies range from non-profit organisations such as the Institute of Electrical and Electronics Engineers (IEEE) and the Internet Engineering Task Force (IETF) to industrial consortia. An example of the latter is the Asynchronous Transfer Mode (ATM) - Forum in which industries strive for consensus on technical issues. To 'qualify' as a grey standards forum, consortia must be open and multi-vendor oriented. The number of such consortia is still growing.[20] It appears to be a phenomenon which is characteristic for the field of ICT. I can think of two main reasons why ICT competitors presently cooperate on a larger scale. Firstly, the field is characterized by multiple players in a growing market. In other words, the good fortune of one player need not occur at the expense of other players. Secondly, compatibility is a (saleable) feature of ICT products and services, and is generally also a prerequisite for new facilities and services. Both motivations for cooperation gain force in market that has started to crystallize.

Grey standards forums vary in their appreciation of close ties with the official standardization circuit. Some opt for cooperation agreements. In general, cooperation involves drafting grey standards and submitting them for formalization to international bodies via national or regional channels. In the past, the IEEE and the European Computer Manufacturer's Association (ECMA), for example, have used their national channels and their liaison relationship with the international standards bodies, respectively, to give their standards a broader base. In 1987 the formal international bodies introduced the Fast-track Procedure by which grey drafts can skip the first stages of the formal process: a specification can immediately enter the process as a Draft International Standard (DIS). To illustrate the improvement, the time required for formalization of an ECMA-DIS was thereby reduced to less than two years.[21] The Publicly Available Specification (PAS)-procedure (1994), which was initiated in response to Internet standardization (IETF), further eased the process of formalization. The IETF, for example, has submitted draft standards to JTC1. Some of its members also participate in the formal process.

Shifts in Formal Standardization

The role that the formal standards bodies play, although proven to be rather stable, has also shown to be susceptible to outside pressure. The previous section indicates that a shift in emphasis has taken place from 'pure standards development' to the inclusion of 'formalization of external standards'; the 'democratic ideals' have been slightly adapted to cater to economic demands for timely standards; and through cooperation agreements the rules for national participation have been circumvented in order to include contributions from regional alliances, social networks (e.g. Internet Society), and formal regional standards bodies. Table 2 summarizes these shifts in emphasis and extrapolates a number of them, based on the pressure for change voiced by participants in formal standardization and my expectation that the 'competition' from grey standardization keeps up. I expect that the formal bodies will ultimately widen their scope and include provisions and procedures for implementation-oriented activities, because the use and implementation of formal standards is a recurrent problem. I further expect that the emphasis in formal standardization will shift from addressing potential interest groups to addressing potential contributors as participants in the process.

COMPARING STYLES OF STANDARDIZATION AND INSTITUTIONAL EFFECTS

Committee standardization is not specifically an innovative activity. Technological knowledge is above all an input characteristic of the standards process and not an output characteristic. Standards codify an amalgam of technical knowledge and a force field of interests. They may lead to standards-compliant innovations and even to innovative technological trajectories, but such statements are hard to substantiate. In part, the lack of clarity about the relation between standardization and technological innovation results from treating *the standards process, standards* and *standards implementations* as one and the same thing, whereas, each may have a distinct effect on technology development. For example, committee standards that are judged as being too complex and expensive to implement — and in this respect have little impact — sometimes focus company R&D or are used as a basis for grey standards (e.g. SGML led to XML).

The distinction also helps to typify different styles of standardization. Roughly

Aspects of formal standardisation:	shift from ...	to ...	and to ...
Priorities	process	outcome	use
Specified priorities	democratic process	timely delivery of standards	implementability of standards
Basis for participation	national	regional	social network
	potential interest groups		potential contributors
Ideological rationale	democratic	economic	
Role of standards bodies	standards development	formalisation of standards	
Scope of activity	standardisation	standardisation, implementation, testing & marketing	

Table 2: Expected shifts in emphasis with regard to formal standardisation.

speaking, the standards process, standards and implementations are successive occurrences in *formal standardization*; in *grey standardization* they are parallel occurrences; and in *de facto standardization* standards follow implementations. Therefore, these styles of standardization affect technology development in a different manner. Furthermore, their source of coercion, their strength, lies in other areas, namely: in the democratic process, in multi-party use of standards and in control of the market, respectively. A last main difference between the three styles of standardization is their concern with implementation-independent standards. The formal standards bodies strive for standards that do not favor certain companies (i.e. highly implementation-independent solutions). In contrast, de facto standards are set in an application environment, and are thus inherently implementation-dependent. The standpoints of grey standardization groups vary greatly on this issue. The consortia will generally favor implementation-independent solutions in order to create equal chances. Internet standardization, on the other hand, addresses a specific implementation environment. Compatibly among Internet standards requires implementation-dependent solutions. The varying degree of implementation-independence implies a different degree of constriction on technology development. The higher the degree of implementation-independence, the less constrictive. See Table 3.

An important problem in generalizing about the nature of the standardization - technology relationship, even when we restrict ourselves to the field of ICT, is that the role of standardization is continuously being redefined. For example, whereas the most common form of standardization used to be *variety reduction ex post*, in the late 1970s and 1980s it was *ex ante standardization* which attracted the attention of standardizers. Standardization in an early phase of technological maturity was seen to be a more effective way to achieve interoperability. It was expected to prevent incompatibility. Since then attention has shifted to acceleration of the process and *ways to deal with multi-party de facto standards* such as Internet standards. Reinterpreting these changes of meaning in terms of interoperability, the problem of interoperability is successively solved through reducing diversity, preventing

style of/aspect in standardisation	formal standardisation	grey standardisation	de facto standardisation
process, standards & implementations	successive occurrences	parallel occurrences	standards follow implementations
source of coercion	democratic process	multi-party use of standards	control of the market
implementation-independence	high	medium	low

Table 3: Characteristics of three styles of standardisation.

diversity and selectively sanctioning diversity in combination with initiating multi-protocol standardization. These solutions embed different views on the role of standardization.

Most economic studies treat standardization as an endogenous factor in *market development* (e.g. David & Greenstein, 1990). Standards mediate the market. This is most evident for assembled and interworking products. The required compatibility standards are points of reference for interdependent market partici-pants. The above noted forms of standardization, that is ex post standardization, implementation independent ex ante standardization, and implementation-ori-ented grey and de facto standardization, affect the market differently. Standardiza-tion activities retrospectively structure the market, accompany preparatory market activity and propel market activity respectively.

DISCUSSION

Those who criticize the formal standards bodies for not delivering innovative and usable ICT standards, are they barking up the wrong tree? An institutional analysis of the standards procedures gives insight into the likely character of negotiation processes (e.g. political consensus in a technical jacket) and hints at the standards that are likely to evolve (e.g. compromises and multiple options) and the ensuing standards implementations (e.g. partial implementations). But, specific effects of institutional procedures on standards outcomes and ensuing technolo-gies are hard to identify. For, the institutional analysis also shows that the role these bodies aim to play is foremost a process-oriented one. Expectations in respect to formal standardization should therefore foremost be directed towards standards process characteristics and not towards standards outcomes. The problem is, of course, that process and outcome are related. The ideological rationale that under-lies institutional provisions is that democratic and impartial working procedures maximize the likelihood that formal standards will be widely implemented. If wide implementation does not occur, one could conclude that the rationale does not work for the field of ICT. For the moment such a conclusion would be unjustified. In rapidly evolving areas such as, for example, mobile telecommunications the formal standards bodies offer an important platform for consensus building. Those who wish to draw an unfavorable comparison between the formal and the grey standards bodies must remember that the latter are 'one issue' bodies, are less concerned with producing implementation-independent standards, and are not burdened with heavily process-oriented aims.

Institutional criticism is usually based on the assumptions, that standardization foremost addresses emergent technologies; that the objective is to produce standards with new technical content; and that the most important stage of standardization is the negotiation stage. With regard to the first two assumptions, past accounts overemphasize the inventive quality of technical contributions. The input of technology suppliers in the formal standards process is not likely to contain true novelty. At stake are premature innovations (i.e. not inventions in the classic sense). These are past the stage of the drawing board but may not yet have a market. I hold that in such cases the standards environment should be viewed as an arena in which the coordination of technology diffusion is being prepared. It should be judged on whether it takes the width of the field into account (i.e. in line with the current, foremost process-oriented aim of formal standardization). In other cases, where innovations are at an early stage of market development, the standards process should be judged on whether standards are implementable and usable, that is, on the standards outcome. The need for such a twofold assessment is a consequence of my view that nowadays committee work foremost aims at creating a common platform of (minimum) requirements for the diffusion of standards uses — and thereby for the diffusion of technologies. The role of formal standards bodies may thus need to be reassessed in terms of its role in coordinating technology diffusion. The emphasis would then lie on standardization as an ex ante market mechanism — and not on standardization as a setting of technological innovation.

ENDNOTES

1. The linear description of technology development denies complex feedback mechanisms (e.g. Pinch & Bijker, 1987; Slaa, 1987). Abernathy & Clark (1985) argue that technological maturity should be coupled to market competition (established/new markets). Fleck (1988) argues that processes of user innovation occur during diffusion and introduces the term *innofusion* to capture this phenomenon.
2. March & Olsen, 1989, p.29. See also Besen, 1990, p.524.
3. The standardization ideology is inferred from reactions of the national standards bodies, CEN, CENELEC, and ISO/IEC on the European Green Paper on standardization (Commission of the European Communities (CEC), 1990), from policy documents of the IEC (IEC, 1992), documents on co-operation (ITU/ISO/IEC, 1993) and secondary literature.
4. CEC, 1991, p.19; Irmer, 1991, p.2.
5. See e.g. Lagerweij 1991a; CEC, 1991, p.21. In the ITU, 'quality' refers to reliability (e.g. Rutkowski, 1991, p.287, p.292) and—especially in the eyes of developing countries—to the usefulness of standards (Codding & Gallegos, 1991, p.361).
6. CEC, 1991, p.36 and CEC, 1990, p.3 respectively. Open membership would hinder a "balanced input" of interests and present difficulties for the effective participation of smaller and medium-sized companies, consumers and smaller countries (CEC, 1991, pp.30-31, p.36).

7. E.g. "interest-balanced" (CEC, 1990, p.3), "broad representation" and "true national consensus" (IEC, 1992, p.12).
8. IEC, 1992, p.9 and p.6 respectively.
9. E.g. CEC, 1991, p.25; IEC, 1992, p.5, p.16.
10. E.g. Bonino & Spring, 1991; CEC, 1991, pp.19-20, p.35; Irmer, 1992.
11. E.g. IEC, 1992, p.5; Lagerweij, 1991b.
12. An example is the idea of publishing the results of pre-standardization activities (IEC, 1992, p.10).
13. E.g. Schmidt & Werle, 1993; Irmer, 1992, p.4. This idea is embedded in the "acknowledged rule of technology, i.e. a technical provision acknowledged by a majority of representative experts as reflecting the state of the art" (ISO/IEC Guide 2, 1991, 1.5).
14. CEC, 1991, p.14; CEC, 1990, p.5; ITU/ISO/IEC, 1993, p.9, p.27.
15. An explicit—though unfortunate—example is given in Schmidt & Werle (1993).The example concerns telefax standardization in the ITU. In short, there were already two standards, namely the Group 3 standards for analogue networks and the Group 4 (OSI-aligned) standards for digital networks, when the adoption of the Group 3 digital telefax standard was proposed. To prevent this from happening, Group 4 proponents came up with a compromise standard. In the ensuing stalemate, the political level of the ITU was called on to make the decision. It decided to accept both proposals. (Schmidt & Werle, 1993, p.26)
16. " (...) In order to solve the problem of membership in respect to ETSI, the definition of a regional standards or standardising body should be modified (...)." (01 (Central Office) 884, July 1992, ISO/Council 1992 - 6.3.1, p.4)
17. E.g. IEC Bulletin, XXV/135, 1992, p.1, p.10.
18. In that period DG III is the Directorate-General for the Internal Market and Industrial (Commission of European Communities) and DG XIII is the Directorate-General for Telecommunications, Information Industries and Innovation (Commission of European Communities)
19. More evidence to support this line of argument is provided in Egyedi (1996, pp.156-160).
20. See e.g. the number of consortia in the field of mobile communications.
21. In comparison, the time JTC1 needed to develop a base standard was on the average almost 4.5 years. (ECMA, 1993, *Computer and Telecommunication Standardization, who's business?*)

REFERENCES

Abernathy, W.J. & K.B. Clark (1985). Innovation: Mapping the winds of creative destruction. *Research Policy 14*, pp.3-22.

Besen, S.M. (1990). The European Telecommunications Standards Institute, A preliminary analysis. *Telecommunications policy*, December 1990, pp.521-530.

Besen, S.M. & J. Farrell (1991). The role of the ITU in standardization: pre-eminence, impotence or rubber stamp? *Telecommunications Policy, 15(4)*, pp.311-322.

Bijker, W.E. (1990). *The Social Construction of Technology*. Dissertation. University of

Twente, the Netherlands.

Bonino, M.J. & M.B. Spring (1991). Standards as change agents in the information technology market. *Computer Standards & Interfaces, 12,* pp.97-107.

Bruins, Th. (1993). *Open systemen.* PTT Research, the Netherlands.

Callon, M. (1986). The Sociology of an Actor-Network: The Case of the Electric Vehicle. In: Callon, M., Law, J. & A. Rip (Eds.). *Mapping the dynamics of science and technology.* London: MacMillan, pp.19-34.

Cargill, C.F. (1989). *Information Technology Standardization. Theory, Process and Organizations.* Digital Press, Digital Equipment Corporation.

Codding, G.A. & D. Gallegos (1991). The ITU's 'federal' structure. *Telecommunications Policy,* pp.351-363.

Commission of the European Communities (October 1990). *Commission Green Paper on the development of European Standardization: Action for faster technological integration in Europe.* Brussels.

Commission of the European Communities (December 1991). *Standardization in the European Economy.* (Follow-up to the Commission Green Paper of October 1990). Communication from the European Commission, Brussels.

Cramer, J. & J. Schot (1990). *Problemen rond innovatie en diffusie van milieutechnologie.* Een onderzoekprogrammeringsstudie vanuit een technologie-dynamica perspectief. Rijswijk: Raad voor het Milieu- en Natuuronderzoek, pp.5-9.

David, P.A. & S. Greenstein (1990). The economics of compatibility standards: an introduction to recent research. *Economics of Innovation and New Technologies, Vol. 1,* pp.3-41.

Egyedi, T.M. (1993). *Torn between values and interests: standardization in a European context.* In: A. Clement, P. Kolm & I. Wagner (Eds.), NetWORKing: Connecting Workers In and Between Organizations. IFIP Transactions A-38, North Holland: Elsevier, pp.181-190.

Egyedi, T.M. (1996). *Shaping Standardization: A study of standards processes and standards policies in the field of telematic services.* Delft, the Netherlands: Delft University Press. Dissertation.

Egyedi, T.M. (1997). Examining the relevance of paradigms to base OSI standardization. *Computer Standards & Interfaces, 18,.* 431-450.

Fleck, J. (1988). *Innofusion or diffusation? The nature of technological development in robotics.* Edinburgh PICT Working Paper, No. 4.

Genschel, Ph. (1993). *Institutioneller Wandel in der Standardisierung van Informationstechnik.* Dissertation. University of Köln, Germany.

Hanseth, O., Monteiro, E., & M. Hatling (1996). Developing Information Infrastructure: The Tension between Standardization and Flexibility. *Science, Technologies and Human Values, 21 (4),* 407-426.

Hughes, T.P. (1987). The evolution of large technological systems. In: W.E. Bijker, Th.P. Hughes & T.J. Pinch (Eds.), *The Social Construction of Technological Systems. New Directions in the Sociology and History of Technology.* Cambridge, Massachusetts: MIT Press, pp.51-82.

IEC (1992). *Masterplan.* IEC strategy for the future.

Irmer, Th. (1992). *Global telecommunications standardization: A reality in the past, but also in the future?* TTC conference November 1992, Kyoto, Japan.

ISO/IEC (1992). *ISO/IEC Directives-Part 1: Procedures for the technical work.* Second edition. Geneva.

ITU/ISO/IEC (1993). *Guide for ITU-TS and ISO/IEC JTC 1 co-operation.* Annex A to WTSC Recommendation A.23, Annex K to ISO/IEC JTC 1 Directives. March 1993.

Lagerweij, M. (1991a). NNI-voorzitter drs. J.C.Blankert: "Normaliseren draait om ordening". *Maatschappijbelangen, 3,* March 1991, pp.38-41.

Lagerweij, M. (1991b). Het resultaat van normalisatie is consensus. *Maatschappijbelangen, 3,* March 1991, pp.42-44.

Mansell, R. & R. Hawkins (1992). Old Roads and New Signposts: Trade Policy Objectives in Telecommunication Standards. In: F. Klaver & P. Slaa (Eds.), *Telecommunication, New Signposts to Old Roads.* Amsterdam: IOS Press, pp.45-54.

March, J.G. & J.P. Olsen (1989). *Rediscovering institutions. The organizational Basis of Politics.* London: Macmillan.

Pinch, T.J. & W.E. Bijker (1984). The social construction of facts and artefacts: or how the sociology of science and the sociology of technology might benefit each other. *Social Studies of Science, 14,* pp.399-441.

Powell, W.W. & P.J. DiMaggio (Eds.)(1991). Introduction. In: W.W. Powell & P.J. DiMaggio (Eds.), *The new institutionalism in organizational analysis.* Chicago: University of Chicago Press, pp.1-40.

Raymond, M. (1990). Divisiveness by design. *Communications International,* May 1990, pp.15-17.

Rutkowski, A.M. (August, 1991). The ITU at the cusp of change. *Telecommunications Policy*, pp.286-297.

Schmidt, S.K. & R. Werle (1993). *Technical Controversy in International Standardization.* Max-Planck-Institut für Gesellschafsforschung, Köln, MPIFG Discussion Paper (93/5).

Schmidt, S.K. & R. Werle (1998). *Co-ordinating Technology. Studies in the International Standardization of Telecommunications.* Cambridge, Mass.: MIT Press.

Slaa, P. (1987). *Telecommunicatie en beleid. De invloed van technologische veranderingen in de telecommunicatie op het beleid van de Nederlandse overheid inzake de PTT.* Dissertation. Free University of Amsterdam. Amsterdam: VU Uitgeverij.

Sweeney, T. & C. Mendler (1994). TCP/IP: It's Official, *Communications International, 27 June 1994,* p.40.

Chapter V

Selected Intellectual Property Issues in Standardization

Martin B. H. Weiss and Michael B. Spring
University of Pittsburgh

INTRODUCTION

This chapter explores several issues related to intellectual property in the standardization process. Different intellectual property issues dominate in the three major phases in the life cycle of a standard. In development, where the content of the standard is created, there are issues related to the intellectual property codified in the standard. During distribution, where documentation on the standard is shared with the vendor and user community, there are issues related to the copyright of the document. In implementation, where products and processes based on the standard are brought to market, there are again issues pertaining to the intellectual property in the standard. Many of these issues have implications for policymakers and stakeholders. The implications are considered and the options available to policymakers and stakeholders are enumerated.

BACKGROUND

Much of the research in standards and the standards setting process has focused either on the technical issues surrounding particular standards or on the economic issues of the standards development process. In this chapter, we consider the intellectual property issues of standards and the standards setting process at all points in the life cycle of a standard. Our discussion may be naive from a legal perspective because we are not lawyers practicing in this area. Nonetheless, we believe that some significant unresolved issues exist at the intersection of intellectual property and the standardization process.

One may view standardization as a process involving three distinct phases[1]. Standardization includes:

1. the development of the standard (*i.e.*, the specification, in an unambiguous and public form, of that aspect of the process or product that is to be the same for all implementations),

2. the distribution of the standard to the vendor and user community, and

3. the implementation of the standard (*i.e.*, the development of products that conform to the standard).

In each phase, distinct intellectual property issues exist. The changing nature of the information technology standardization process introduces some further intellectual property issues that must be considered[2].

Traditionally, in the standards specification phase, standards formalized and made public an existing dominant industry practice or technology. The ownership of protected intellectual property related to the standard was clear, most often in terms of a patent. When a standard is to be a US national standard, in accord with the Procedures for the Development and Coordination of American National Standards, the owners of any intellectual property rights agree to license the product at a fair price and in a nondiscriminatory way (ANSI, 1987, p. 28). In contrast to this historical tradition of standards sanctioning an existing well-defined product, standards today may precede products (Cargill 1989, Weiss 1991a, Bonino and Spring 1991)[3]. Technologies can be developed in committee during the development of the standard, leaving the disposition of intellectual property rights uncertain.

In the development of a standard, significant issues revolve around the ownership of intellectual property. The relevant intellectual properties are often covered by patents and copyrights on the contributions of the committee. While the software patent tradition is still evolving, patents can apply to software algorithms. For hardware-based contributions, the patent protection is much more clearly applicable. Similarly, copyright law can also apply if the contributed software has been protected by its owners. This protection would only apply to the actual code, not the underlying algorithm, in accordance with traditional interpretations of the law (OTA92b). Recent changes in the policy of the Patent and Trademark Office allow patents to be issued on some types of algorithms (Samuelson 1992, Chisum and Jacobs 1992). Modifications to contributed software could also be protected, as they would be considered derivative works. Trade secret law does not apply because when a technology is brought into the committee, it is no longer a secret, but rather a matter of public record. Since some intellectual properties are developed in the public, voluntary committees, questions are raised about who owns the ideas.

In the dissemination phase of standardization, the issues pertain to the copyright of standards documents. Today, copyright is almost always owned by the standards development organization (SDO) sanctioning the activity. For a number of SDOs, revenues from the sale of standards have become significant. As technologies converge, diverse SDOs must harmonize their standards with other SDOs. In many cases virtually the same intellectual property (*i.e.*, standard) is sold under multiple different names. This competition, and the resulting revenue implications, has caused some contention to arise between SDOs (OTA 1992a).

Finally, there are intellectual property issues in the implementation phase of standards as vendors build products that are based upon the standard. With the emergence of anticipatory standards and complex highly interdependent standards, the development of tests to ensure product compliance becomes more important. Since products developed to specifications set forth in standards sometimes fail to interoperate, there has been increasing pressure to develop unambiguous languages for the specification of standards that will allow for an automated development of test suites. When test suites are developed by an organization other than the SDO, one may ask whether the tests are derivative works of the standard and therefore protected under the copyright statutes[4], much as computer executable code is a derivative work of computer source code.

Forms of Intellectual Property Protection

Historically, the state has protected intellectual property (IP) because this protection (presumably) meets social objectives. That is, the IP protections should be designed such that the specific social goals are advanced by that protection (OTA 1986, Ch. 3). In this paper, we assume that IP protections are beneficial and meet social goals. IP protections in general are embodied in the US constitution, and are designed to foster the dissemination of information and to "promote the useful arts" (Chisum and Jacobs 1992). Four basic mechanisms exist for the protection of intellectual property: patents, copyrights, trademarks, and trade secrets[5]. Of these, trademarks and trade secrets are generally not relevant to this discussion.

It is generally agreed that the purpose for intellectual property protection (via patents and copyrights) is to motivate individuals and firms to produce such property. Intellectual property laws provide a monopoly on an idea or an expression that would not be possible by other means. Were it not for these forms of protection, an idea or expression, once conceived, could not be placed into the public domain without the total loss of benefits to the developer of the idea or expression. Under these circumstances, individuals would lack an incentive to produce more ideas (or at least to place them into the public domain). In terms of this incentive one may argue that any intellectual properties associated with compatibility standards, which must by definition be public, are absolutely dependent on the availability of intellectual property protections.

Patents are issued to
"[w]hoever invents or discovers any new and useful process, manufacture, or composition of matter, or any new and useful improvement thereof . . . (35 USC 101)

	Copyright	Patent
Person	• Author/Work for Hire	• Inventor
Subject Matter	• Originality • Fixed in Medium of Expression • Non-Utilitarian	• New • Non-Obvious • Utility
Procedure	• Automatic • Registration to preserve legal remedies	• Must be applied for and approved/granted

Table 1: Copyrights vs. *Patents*

In contrast, copyrights are issued for:

"works of authorship . . . fixed in any tangible medium of expression, now known or later developed, from which they can be perceived, reproduced, or otherwise communicated, either directly or with the aid of a machine or device (17 USC 102(a)).

Loosely, one could state that patents cover the works of inventors and copyrights cover the works of authors. Table I illustrates the differences between patents and copyrights from a subject matter and procedural perspective[6].

These statements from the law ignore the rich legal history that exists around the patentable and copyrightable subject matter. The case that is often cited that draws a distinction between patentable subject matter and copyrightable subject matter is *Baker vs. Selden*[7]. In this case, Selden developed and published a business bookkeeping system that made use of a series of forms and explanatory essays. When Baker published a book containing essentially the same forms, albeit with different explanatory essays, Selden sued for copyright infringement. Although lower courts supported Selden's position, the Supreme Court overturned the verdict and ruled in favor of Baker (Samuelson 1992). Chisum and Jacobs (1992) identifies three propositions for copyright law that emerged from this case:

Baker copied only the idea, not the expression. While the forms were essentially identical, their explanations differed.

". . . a person may copy even the expression of the author 'for the purpose of practical application' rather than for the purpose of 'explanation'"

The bookkeeping forms were not an author's "writings", hence they were not copyrightable subject matter.

As many information technology standards are international in nature, international agreements on patents and copyrights are of interest. The bilateral, regional, and international reciprocity agreements generally provide, with a couple notable exceptions, reasonable copyright protection. The Berne Convention for the Protection of Literary and Artistic Works of 1886 and the Universal Copyright Convention of 1952 provide for certain minimum protections using the national treatment rule, *i.e.*, the work of a foreign author is protected in a given country in the same way the work of a native author would be protected. More than 70 nations are signatories of the two conventions (Nordheim *et.al.* 1990, Stewart 1989). While questions about protection of copyright remain, they are essentially moot within the set of nations concerned with information technology standards and technology, as most of these countries are signatories to one or both of the multilateral conventions discussed above.

INTELLECTUAL PROPERTY ISSUES IN THE PROCESS OF STANDARDIZATION

Numerous researchers have studied the process of standardization. In particular, Weiss and Cargill(1992), Lehr(1992), Bonino and Spring(1991), Weiss(1991), and Cargill(1989) have examined some of the ways in which the process of standardization is changing in the information technology arena. The issues here involve the globalization of standards, the emergence of consortia, the anticipatory nature of selected standards, the complexity and interdependence of standards, etc. In the next three sections, intellectual property issues pertaining to the changing nature of information technology standards and standardization are discussed in terms of the major phases of standardization — development, dissemination, and implementation. Three significant changes in the standardization process serve to focus the intellectual property issues explored[8]:

1. In the development of anticipatory standards, intellectual properties are more likely to be created in the standards committee.
2. Regarding the dissemination of standards, the publication revenues from standards have become important contributors to the financial well-being of some standards development organizations.
3. Implementing standards is a matter of assuring product conformance. Increasingly, conformance tests and other conformance mechanisms are tightly coupled to the standards.

The Development of Standards

Historically the development of a standard has involved vendors bringing technology to the standards committee so that it can be considered for inclusion in the standard. The vendors compete with each other because the vendor whose approach or technology serves as a basis for the standard may have a competitive advantage in the subsequent marketplace for standardized products. In this case, the disposition of intellectual property is fairly clear. The technology is owned by the vendor; they developed it and brought it into the committee. If it is included in the standard, the vendor has the right to license the technology, and, under ANSI rules (ANSI, 1987, p. 28)[9], the obligation to do so without discrimination and "at reasonable cost". Note that these rules apply specifically to ANSI accredited US standards bodies. Other SDOs, such as nonaccredited committees and non-US committees, may have different rules. In the case of non-US committees, the owners of the IP, granted through ANSI, would be protected under existing international agreements.

While the process of developing standards based on existing stable products assures stable standards, a different process is required when product life cycles are shorter than the standards development life cycle. The standardization process is time consuming, particularly if the number of participants is high and they have divergent preferences (Weiss 1991b). When a market exhibits rapid technological growth, as the information technology market has, the time required to develop a standard may be longer than the product life cycle. To cope with this, standards bodies have begun to act in anticipation of the technology, developing standards before products are produced. These are known as anticipatory standards (Cargill 1988). In fact, Bonino and Spring (1991) argue that anticipatory standards act as mechanisms for collective market planning; *i.e.*, they are an embodiment of a private industrial policy.

With anticipatory standards, the model of standards development is quite different. In order for anticipatory standards to be an effective method for standards development, the industry requires something of a unified vision of the future[10]. In other words, the system environments of the future that are likely to be dominant have to be articulated, and the necessary standards infrastructure understood and defined. This imposes a substantial strategic planning requirement on the SDOs and the industry as a whole. In contrast to anticipatory standards, consortia have also emerged that attempt to achieve the same goal by speeding up the standards development process (Weiss and Cargill, 1992).

New ideas are being developed in, and as part of, the standardization process, however it is achieved. While organizations may still submit technologies to the committee, for inclusion into the standard, substantial portions of the technology may be either motivated by the committee or actually developed in the committee. The intra-committee development may be the result of a compromise that was necessary for the standards development process to proceed. If the compromise was based on the adoption of a single technology, the intellectual property issues are similar to the traditional case. If the compromise was based on the merger of two proprietary technologies, then the ownership of the resulting technology is less clear. Finally, if the compromise resulted in the development of a third alternative technology not encompassing either of the original technologies, then it is completely unclear who owns the rights to the technology.

Similar problems are faced by consortia developing standards, although it is easier to dispose of the intellectual property issues, because the consortia are essentially private organizations working to further the benefit of their members.

Dissemination of Standards

The SDO almost always holds the copyrights on standards documents and is in a position to determine how the documents will be disseminated as well as how any revenues that accrue from the sale of the documents will be allocated. While the dissemination of standards is generally managed directly by the standards development organization, this responsibility may be delegated to a publisher or other party. Thus, the SDO normally controls dissemination of the standard by directly or indirectly publishing and selling the standard to interested parties. In many cases the SDO derives significant revenue from the sale of standards and carefully guards its copyright so as to maximize revenues. Other SDOs subcontract what may be a break-even activity to an organization that can do the publishing on a more cost-effective basis. One of the more uncommon publishing policies is that of the Internet Engineering Task Force (IETF). The IETF is responsible for developing and maintaining the standards used on the Internet, such as TCP/IP, SMTP, *etc.* The IETF makes their standards, called RFC's, available electronically, and waives all publishing or use fees. A common argument in the telecommunications research community is that this policy may have helped TCP/IP in attaining a significant share of the market. The argument is that since university researchers could obtain the standards at no (direct) cost in electronic form they elected to choose these standards over CCITT or ISO standards[11]. In fact, this has become a very important issue of late, as articulated in the OTA report on standards (OTA 1992a).

We contend that information technology standards, even more than other technical compatibility standards, need to be more widely disseminated. This is because users as well as manufacturers must often have significant knowledge about the standards on which their systems are based if they are to use them efficiently and effectively. For example, users did not really have to understand the details of paper and typewriters in the past, whereas they do have to understand electronic document processing standards such as SGML, ODA, and EDI[12]. More concretely, IBM doesn't need to know about airplane standards, even though they use them frequently when IBM employees fly or when equipment is shipped via air mail, whereas Boeing needs to know about information technology standards to operate efficiently and effectively as a manufacturer of aircraft. Thus, the market for information technology standards is relatively large and this has significant financial implications for both producers and consumers of standards documents.

Because of this market expansion, the revenues generated by standards documents have become more significant to the SDOs, to the point where they are a significant fraction of the overall revenues of the SDO. With this much revenue at stake, the SDOs have become very contentious, as documented in the OTA report (OTA 1992b). Indeed, there is some suggestion that the pricing of standards by some SDOs may be negatively effecting their adoption. Interestingly, originators of standards may choose to shop among SDOs, just as authors shop among publishers, until they find the SDO that will provide for the widest possible dissemination of their ideas[13].

Implementation of Standards

Once the standards development is complete and the document has been published, it remains for vendors to develop products. David's taxonomy of standards would characterize most information technology standards as technical compatibility standards (David, 1987). Compatibility, or conformance to the standard, may be measured either of two basic ways:

1. The conforming product must interoperate with other products that are known to conform. When standards are relatively simple and based on existing products, it is possible to define conformance in terms of actual interoperation. In fact, with hardware standards, semiconductor manufacturers built integrated circuits that embodied the standard. As long as these IC's were compatible, it was likely that the products would also be compatible.

2. The conforming product must pass some conformance test(s). As standards become more complex, more option based and are developed in advance of products, the conforming products must pass a series of tests to assure compatibility.

Historically, vendors have developed conformance tests for standards that have been applied internally. As the costs of incompatible systems have grown, major users and SDOs have begun to call for third party certification of conformance. The conformance tests are designed to determine whether a particular product conforms to the standard. Thus, conformance tests are used by vendors and users as a predictor of interoperability, which is the user's primary concern[14]. If a product passes the conformance test, it is believed to be likely to interoperate with other products that also passed the conformance test[15].

There has been increasing pressure to develop standards that are absolutely unambiguous. In the end, if a standard is correctly developed, the development of the conformance test should be mechanical. To achieve this kind of precision, a protocol is formally specified in a standard language such as LOTOS or Estelle, and conformance tests are written in Tree Tabular Combined Notation (TTCN). TTCN acts as a high level language to describe the details of the testing procedures. The actual test sequences and low level software that are implemented in the testers are often generated automatically using a TTCN compiler (Linn 1989).

In this case, the development of conformance tests are dependent on the standard. Since conformance tests are simply a more formal expression of the underlying standard, are conformance tests to be considered derivative works of the standard? Current copyright law would suggest that the test sequences are derivative works of the TTCN notation, just as machine code is a derivative work of the original higher level language code (OTA 1992b). Are manufacturers of test equipment liable for royalty payments to SDOs for the use of the test sequences based on TTCN-based conformance tests? If so, is some royalty due the SDO that owns the copyright to the standard?

COPYRIGHT PROTECTION FOR STANDARDS DOCUMENTS

At the current time, any patent-related fees accrue to the organization that owns and contributed the technology to the standard. Document royalties, if there are any, normally accrue to the SDOs. It is important to ask several questions related to this current fee structure: Are the costs for standards development allocated fairly throughout society? What is the basis for IP ownership? Should any changes be made? Of these, we will focus on the basis for, and strength of, copyright protection of standards documents.

Basis for Copyright Ownership

The work of developing a standard is performed by volunteers (from the SDO's point of view). These volunteers meet periodically to discuss the technical content and to agree on the text of the document that will ultimately become the standard. The SDO provides some support services to these "volunteers", but does not participate actively in the preparation of the document. Thus, the SDO cannot legitimately claim creative contributions to the standard. The bases for IP ownership by the SDO are:

- that the standards are considered "works for hire," hence the SDO owns the work;
- the committee members and the SDOs are engaged in "joint work," in which case both the SDO and the committee participants own the copyright;
- the work is considered a "compilation" by the SDO, in which case the copyright belongs to the SDO as the assembler of the component parts of the compilation (*i.e.*, the standard); or
- the committee participants collectively assign the rights to the IP (particularly the copyright of the standard itself) to the SDO.

The "works for hire" and "compilation" paradigms are perhaps the most solid bases. The Copyright Law (17 USC 101) states that a work for hire is either created by an employee within the scope of employment or a work that is specially commissioned or ordered. When the SDO proposes development of a new standard, it could be said that they are commissioning a work. The code does not specify the nature of the contract, so the fact that the contributors to the standard are volunteers may be irrelevant. Weiss and Sirbu (1991) have argued that a standard consists of a collection of technologies. Thus, there is some basis to consider that a standard is, indeed, a compilation. In this case, the compiler owns the copyright of the compilation, even though the individual contributors own the copyright on their contributions (17 USC 201)[16]. Since it was the committee that performed the compilation, not the SDO, it is unclear that this is an appropriate basis under the law.

Scope of Copyright Protection

In this section, we will examine copyright traditions in the US as they are relevant to standards. Then, we will examine two standards in some detail with respect to these copyright principles. Finally, we will consider the scope of protection for these standards.

General Copyright Concepts Relevant to Standards

If we stipulate that the copyright for a standards document can be legitimately owned by the SDO, then it is also appropriate to consider the extent to which the copyright laws would protect the standard. A standards document is a technical document that states the facts, specifications, algorithms, *etc.* necessary to implement the standard. The courts have traditionally granted relatively limited protection to works of this nature.

In *Feist Publications Inc. vs. Rural Telephone Service Co.*[17], the Supreme Court held that a simple compilation of facts (such as the telephone directory involved in this case) is not copyrightable due to the lack of originality. While facts themselves are not copyrightable, the compilation of facts may be — provided that the arrangement, display, selection, *etc.* are original or are creative. Even in this case, the copyright protection is extended only to the arrangement, display, selection, *etc.*, not to the underlying facts (Chisum and Jacobs 1992, p. 4-14). Furthermore, in *Baker vs. Selden* (discussed above) the Supreme Court clearly articulated that ideas, such as those represented in Selden's forms, were not copyrightable.

Finally, a number of cases exist (see Chisum and Jacobs 1992, pp. 4-26 to 4-30) that fail to extend copyright protection to documents in which the idea and expression of the idea are said to merge. This "merger doctrine" of copyright is said to hold when a very limited number of ways exist to express an idea. In *Lotus vs. Paperback*, Judge Keaton found that the idea of the "rotated L" spreadsheet display, where the column labels are displayed on the top of the screen and the row labels are displayed on the left of the screen, was an example of such a merger (Samuelson 1992).

Examination of Particular Standards

To consider the scope of copyright protection available for a standards document, let us consider some examples of standards documents. The examples we will use are CCITT Recommendations Q.920 and Q.921 (Digital Subscriber

Signalling System No. 1, Data Link Layer). The tables of contents for these standards are listed below. Let us consider each standard from the standpoint of the scope of copyright protection.

Q.920: ISDN user-network interface data link layer - General aspects

1. General
2. Concepts and terminology
3. Overview description of LAPD functions and procedures
    3.1 General
    3.2 Unacknowledged Operation
    3.3 Acknowledged Operation
    3.4 Establishment of Information Transfer Modes
4. Service Characteristics
    4.1 General
    4.2 Services provided to layer 3
    4.3 Services provided to layer management
    4.4 Administrative services
    4.5 Model of the data link service
    4.6 Services required from the physical layer
5. Data link layer - Management structure
    5.1 Data link procedure
    5.2 Multiplex procedure
    5.3 Structure of the data link procedure
Reference

Contents of CCITT Recommendations Q.920

Recommendation Q.920

Recommendation Q.920 describes the Link Access Procedure for the D channel (LAPD) in general terms. It discusses the purpose of LAPD, the manner in which it interacts with other protocols at different layers of the Open Systems Interconnection (OSI) Reference Model, the functions it performs, the services it provides, and how it is managed. Much of the text is descriptive and explanatory, designed to provide a general context for the existence and operation of the protocol. Some detailed technical information regarding the design of the protocol is provided (for example, CCITT 1988, p. 16).

Recommendation Q.921

It is in this standard that the LAPD protocol is specified in significant technical detail. A computer scientist or engineer would be able to implement the protocol with the information contained in Q.921, something which could not be said for Q.920. Recommendation Q.921 specifies the format of the data frame, the size of

Q.921: ISDN user-network interface - Data link layer specification

1. General
2. Frame structure for peer-to-peer communication
    2.1   General
    2.2   Flag sequence
    2.3   Address field
    2.4   Control field
    2.5   Information field
    2.6   Transparency
    2.7   Frame Check Sequence field
    2.8   Format convention
    2.9   Invalid frames
    2.10 Frame abort
3. Elements of procedures and formats of fields for data link layer peer-to-peer communication
    3.1   General
    3.2   Address field format
    3.3   Address field variables
    3.4   Control field formats
    3.5   Control field parameters and associated state variables
    3.6   Frame types
4. Elements for layer-to-layer communication
    4.1   General
    4.2   Primitive procedures
5. Definition of the peer-to-peer procedures of the data link layer
    5.1   Procedure for the user of the P/F bit
    5.2   Procedures for unacknowledged information transfer
    5.3   Terminal endpoint identifier management procedures
    5.4   Automatic negotiation of data link layer parameters
    5.5   Procedures for establishment and release of multiple frame operation
    5.6   Procedures for information
    5.7   Re-establishment of multiple frame operation
    5.8   Exception condition reporting and recovery
    5.9   List of system parameters
    5.10 Data link layer monitor fuction

Annex A -   Provision of point-to-point signal-ling connections
Annex B -   SDL for point-to-point procedures
Annex C -   SDL representation of the broadcast procedures
Annex D -   State transition tables for point-to-point procedures for the data link layer
Appendix I   Retransmission of REJ response frames
Appendix II Occurrence of MDL-ERROR-INDICATION within the basic states and actions to be taken by the management entity
Appendix III Optional basic access deactivation procedures
Appendix IV Automatic negotiation of data link layer parameters

Abbreviations and Acronyms used in Recommendation Q.921

each information field within the data frame, the manner in which error check sequences are to be computed, *etc.* In general, this recommendation contains a great deal of factual information about the operation of the protocol, with very little explanatory text. Much of the recommendation is in the form of figures and tables designed to specify the protocol in detail. The Annexes to Q.921 contain a description of the protocol in State Description Language (SDL) and state transition tables, which together are a very precise way of specifying the protocol.

Discussion

Since very little explanatory text is provided with Recommendation Q.921 beyond what is necessary to describe the LAPD protocol in technical terms, the protection of that standard is quite weak. This comes primarily from the merger doctrine, described above. In particular, the figures illustrating the structure of the protocol, the state transition tables, and the SDL figures, may not be copyrightable at all, because they are factual definitions of the protocol. The accompanying text would be subject to some protection. It could be argued that very few ways exist in which this protocol could be described, and that therefore it is the statement of an idea - a standard protocol that is to be implemented by engineers to enable the interoperation of systems. What may not fall under this heading is the manner in which the standard is organized. Other methods for structuring the information exist beyond the way chosen by the CCITT.

Copyright protection for the Q.920 standard would likely be stronger. This standard is much more explanatory. It would be much more difficult to argue that the merger doctrine would apply to this recommendation.

In the context of the *Baker vs. Selden* case, it would be possible to approximate the forms that Selden used for his business system with the figures and tables in Recommendation Q.921. While the text in neither Q.920 nor Q.921 is particularly explanatory, direct copying of these texts would likely be considered copyright infringement.

DISCUSSION

In the development of any public policy, it is necessary to balance the needs of society against the rights of the members. Stiglitz (1991) has argued that intellectual property protection is incomplete and imperfect, but that this imperfection is not always bad for society. In fact, he argues that:

> . . . intellectual property protection is important, but, almost of necessity, imperfect. It is harder and therefore more costly to define than conventional property, and by the same token, it is costlier to enforce intellectual property rights. In some cases, the inventor gets more than his marginal contribution, in other cases, less. Intellectual property protection strengthens dynamic efficiency and competition, but often at the expense of static efficiency and competition. If overly strong, it can actually hinder both dynamic and static efficiency and competition. Public policy towards intellectual property must take into account this perspective. There is no simple prescription; as in other areas of what economists refer to "the economics of second best", appropriate policy needs to take into account

the facts and circumstances pertinent to different situations.

Thus, we are reminded that the intellectual property rights related to standards must be considered in a social context.

Standards may be developed to provide basic measures, assure quality, or provide for compatibility (David 1987). Information technology standards are predominantly developed to allow for technical compatibility. A compatibility standard must be widely implemented to be successful. Although Katz and Shapiro (1986) point out that this "lock-in" is not always optimal, widespread adoption remains the most important indicator of the success of a compatibility standard.

No definitive list of factors that contribute to the adoption of a compatibility standard has been identified in the literature. Sirbu and Stewart (1986) identified the locus of decisionmaking as important predictors for which product type (*i.e.*, modems) is likely to be subject to standardization, this does not extend to a standard-by-standard comparison within a product type. Weiss and Sirbu (1990) have considered attributes that may indicate technical superiority of a technology in a standards committee, but that is only one dimension and is at too low a level to be useful here. Thus, we are left to propose a set of factors based on informal observation of the industry and general familiarity with the literature. We suggest that the goal of adoption is generally achieved when the standard is[18]:

Accessible: Standards must be accessible to potential users and vendors. A standard that is not accessible is not likely to be widely implemented. As we stated earlier, a theme in the research community hypothesizes that the success of the TCP/IP protocols is largely due to the fact that the standards were very accessible to the academic community, who developed systems (*eg.*, BSD Unix) and applications (*eg.*, electronic mail, distributed file systems) based on that suite rather than the OSI protocols that were emerging at that time.

Timely: Historically, it has been felt that standardization too early in the development process stifled the emergence of the optimal product. More recently, in the information technology arena, the late emergence of standards has resulted in lack of compatibility[19].

Appropriate: As with any product, the appropriateness of the technology to the user's needs and its manufacturability at reasonable cost is important. If a standard embodies technology that is too advanced, then compatible products will be either too costly or late to market. This was the case with the early V.32 modems. These modems have been on the market for approximately five years, and only recently have prices become sufficiently attractive for a mass market to develop. Similarly, a product could be too simple, such that it does not perform the range of functions required by the end users of the technology.

Well Written: A well-written standard is unambiguous, easy to comprehend, concise, and complete. Failure to meet these criteria results in a standard that is open to interpretations due to incompleteness or ambiguity, which results in potentially incompatible implementations.

Fair: Fair, in the sense used here, means that the standards setting process is not biased *a priori* to dominant firms, so that new entrants have an equal voice. See Lehr (Lehr 1992) for a more complete discussion of this.

We equate the motivation of payment for intellectual property, via fees and royalties, with the ownership of intellectual property. At the current time, any patent-related fees accrue to the organization that owns and contributed the technology to the standard. Document royalties, if there are any, normally accrue to the SDOs. It is important to ask several questions related to this current fee structure: Are the costs for standards development allocated fairly throughout society? What is the basis for IP ownership? Should any changes be made?

The costlier a product is to develop and implement, the costlier the product will be in the marketplace. If the product development costs are increased due to royalty payments to the SDOs on all forms of IP, then the social benefit of the standard is reduced. Thus, from a market perspective, it is critical that costs be minimized. Additionally, in accordance with the public goods literature, the costs should be allocated such that those who benefit the most from this public good (the standard) should pay the most.

CONCLUSIONS

When considered from the viewpoint of the changing life cycle of information technology standards, new intellectual property issues appear to the research community. We are not legal scholars, so we are unable to argue the details of current intellectual property law and its interpretation. As standards researchers, however, we are able to discuss the consequences of some of these issues.

As pointed out in the OTA report (OTA 1992a), significant issues face standards developers today. In addition to the issues documented in that report are the issues with respect to intellectual property that have been identified in this discussion.

ENDNOTES

1. While the phases are conceptually distinct, in practice, the phases overlap. Most commonly, the third phase may occur concurrently with the first two phases.
2. There is some evidence that the standardization process is changing across industries in the same way it is in the information technology industry, although the change is most dramatic in the high tech and information technology industries.
3. Standards that are developed and disseminated before compliant products have achieved a dominant market penetration are commonly referred to as anticipatory standards.
4. Indeed, one might ask whether every use of the test suite should involve a royalty payment similar to that paid by artists for the right to perform a given piece of music.
5. This discussion on forms of intellectual property protection draws from the recent OTA report *Finding a Balance: Computer Software, Intellectual Property, and the Challenge of Technological Change* (OTA 1992b).
6. Prof. Pamela Samuelson, Private Conversation.
7. Baker v. Seldent, 101 US 99 (1879).

8. There are other issues, related to but outside the framework of intellectual properties in the process of standardization, that are outside the scope of this paper. For example, this paper will discuss the notion of standards based conformance tests as derivative works under intellectual property laws. It will not discuss whether the developers of standards have liability for tests derived from their standards.

9. In general, license fees are not beneficial to the standardization process if the cost is passed on to the consumers of the product. In this case, higher product prices may limit the adoption of the standard by consumers as well as producers.

10. There is an ongoing debate about the value of anticipatory standards. Since they have to be developed for a future environment, they have to predict markets in the future. If the future is incorrectly anticipated, the standards may be of little or no value. The cost of suites of standards such as ISDN and OSI is high. Some would argue that the cost is not justified by the need to redefine these standards prior to using them.

11. There may actually be non-trivial indirect costs associated with accessing these standards, such as access to computer equipment and attachment to the network.

12. This may simply be a temporary anomaly caused by the relatively recent emergence of these standards.

13. Vendors may also select their SDO based on the likelihood that their preferred technology will be adopted (Lehr 1991).

14. This has lead to the observation by several vendors that they build products to the conformance test, not to the standard.

15. This is by no means a *guarantee* of interoperation, however (Castro 1990). Because of the complexity of the standards it is possible that two products that pass the same conformance tests do not interoperate in a satisfactory manner.

16. Under the 1976 Copyright Law, copyright ownership is automatically granted when the work is originated. If the owner wishes to defend their property rights in court, the work must be registered with the Copyright Office in the US.

17. 111 S. Ct. 1282, 18 U.S.P.Q.2d 1275 (1991)

18. At minimum, this list provides a useful starting point for a more careful discussion of how one identifies a "good" or "successful" standard.

19. Many have argued that both the ISDN and OSI standards are no longer useful or appropriate due to their tardiness. It is very possible that these standards, on which millions and perhaps billions of dollars have been spent by users, vendors, and others, will never be a significant market force.

REFERENCES

ANSI, "Procedures for the Development and Coordination of American National Standards", New York: ANSI, 1987.

Besen, Stanley M. (1986) "Private Copying, Reproduction Costs, and the Supply of Intellectual Property" *Information Economics and Policy*, V. 2, No. 1, pp. 5-22.

Bonino, Michal J. and Michael B. Spring (1991) "Standards as change agents in the information technology market" *Computer Standards and Interfaces*, V. 12, No. 2, September, pp. 97-108.

Braunstein, Yale M. (1989) "Economics of Intellectual Property Rights in the International Arena" *Journal of the American Society for Information Science*, Vol.40, No. 1, pp. 12-16.

Cargill, Carl F. *Information Technology Standardization: Theory, Process, and Organizations* Bedford, MA: Digital Press, 1989.

CCITT (1988) *Digital Subscriber Signalling Systems No. 1 (DSS 1), Data Link Layer* Recommendations Q.920-Q.921, Volume VI, Fascicle VI.10, Geneva 1989.

Castro, Stephen (1991) "The relationship between conformance testing of and interoperability between OSI systems" *Computer Standards and Interfaces* Vol. 12, pp. 3-11.

Chisum, Donald S. and Michael A. Jacobs *Understanding Intellectual Property Law* New York: Matthew Bender 1992.

David, Paul A. (1987) "Some New Standards for the Economics of Standardization in the Information Age", in P. Dasgupta and P. Stoneman (eds.) *Economic Policy and Technological Performance*, Cambridge: Cambridge University Press, 1987.

Farrell, Joseph (1989) "Standardization and Intellectual Property" *Jurimetrics Journal*, Vol. 30, No. 1, pp. 35-50.

Katz, Michael L. and Carl Shapiro (1986) "Technology Adoption in the Presence of Network Externalities" *Journal of Political Economy* Vol. 94, No. 4, August 1986, pp. 822-841.

Linn, Richard J., Jr. (1989) "Conformance Evaluation Methodology and Protocol Testing" *IEEE Journal on Selected Areas in Communications*, Vol. 7, No. 7, September, pp. 1143-1158.

Miller, Arthur R. and Michael H. Davis (1990) *Intellectual Property: Patents, Trademarks, Copyright* West Publishing.

Nordemann, Wilhelm, Kai Vinck, Paul W. Hertin, and Gerald Meyer (1990)*International Copyright and Neighboring Rights Law* Weinheim Germany: VCH.

Office of Technology Assessment (1986), US Congress *Intellectual Property Rights in an Age of Electronics and Information* OTA-CIT-302, Melbourne FL: Kreiger Publishing Co.

Office of Technology Assessment (1992a), US Congress *Global Standards: Building Blocks for the Future*, OTA-TCT-512, Washington DC: US Government Printing Office.

Office of Technology Assessment (1992b), US Congress *Finding a Balance: Computer Software, Intellectual Property, and the Challenge of Technological Change* OTA-TCT-527, Washington DC: US Government Printing Office.

Rutkowski, Anthony M. (1991), "Networking the Telecom Standards Bodies: Version 3.1 (Final), August 2, 1991, IETF Bulletin Board.

Samuelson, Pamela (1992) "Computer Programs, User Interfaces, and Section 102(b) of the Copyright Act of 1976: A Critique of *Lotus v. Paperback*" *Law and Contemporary Problems*, Vol. 55, No. 2, Spring 1992, pp. 311-353.

Sirbu, Marvin and Steven Stewart (1986) *Market Structures and the Emergence of Standards: A Test in the Modem Market* WP-8, MIT Research Program on Communications Policy, June 1986.

Stewart, S.M. (1989) *International Copyright and Neighboring Rights (2nd ed.)* London:

Butterworth & Co.

Stiglitz, Joseph E. (1991) "Public Policy Towards Intellectual Property" *International Computer Law Adviser*, June, pp. 4-7.

Weiss, Martin B.H. and Marvin A. Sirbu (1990) "Technological Choice in Voluntary Standards Committees: An Empirical Analysis" *Economics of Innovation and New Technology* Vol. 1, No. 1, pp. 111-134.

Weiss, Martin B.H. (1991a) "Compatibility standards and product development strategies: A review of data modem developments" *Computer Standards and Interfaces*, V. 12: 109-121.

Weiss, Martin B.H. (1991b) "Standards Development: A View from Political Theory" University of Pittsburgh, Department of Information Science Working Paper LIS042/DIS 91010, June 1991.

Weiss, Martin B.H. and Carl Cargill (1992) "Consortia in the Standards Setting Process" *Journal of the American Society for Information Science*, 43(8), September, 1992.

Chapter VI

Standardization and Intellectual Property Rights: Conflicts between innovation and diffusion in new telecommunications systems

Eric J. Iversen
The STEP-Group

INTRODUCTION

In today's environment of rapidly evolving information and communication technologies (ICTs), technical standardization is said to be confronted by a "minefield" of intellectual property rights (IPRs). Patents and other industrial IPRs that might belong to individual developers of technology have the potential to undermine the collective pursuit of technical standardization that might serve the common interests of the sector or industry. This tension between the individual and the collective, between the development of technology and its diffusion, is however by no means new; it is an inherent feature of standard development as an institution of innovation.

The fact that this tension has only recently been converted into conflict raises a host of interesting questions about standardization in the evolving environment of the 'digital age'. In this chapter, we will address some of these. We are especially interested in the fundamental question concerning the roles of standard development organizations and IPRs in the "technology infrastructure" (Tassey, 1995) and how these roles are "co-evolving" (Nelson, 1995) with the rapidly developing ICT industry. The contention is that this process of coevolution is bringing what are initially complementary functions in the innovation process into increased confrontation.

In this chapter such questions will be explored in terms of innovation-theory in which the role this 'technology infrastructure' plays is explicitly recognized. The discussion of this relationship moreover will be largely presented in terms of a case study, featuring the controversy that arose during the standardization of the now popular GSM system, produced by the European Telecommunication Standards Institute (ETSI).

BACKGROUND

The 'minefield' of intellectual property rights said to face SDOs such as ETSI includes a wide variety of different hazards. There are different types of mines, in the form of patents and more recently copyrights covering software. But more importantly, the different types of IPRs in the minefield pose different types of danger to the standardization process depending especially on their number and the disposition of the owners.

In this chapter, we are interested in what are termed "essential IPRs", as it is this type of mine that is dangerous to standards-development. The potential for conflict arises when the implementation of a standard, by its *essence*, necessitates the application of proprietary technology. A standards development organization, for example, that sets to work codifying a set of standard signal-transmission specifications for a mobile-communication system, will be working in an area where private companies have already researched and perhaps have developed their own solutions. The standards body thus risks infringing the rights of those companies who have invested valuable R&D resources in this field if the standard 'specs' it proposes, by their depth and detail, necessitate the usage of technical solutions that are protected by active IPRs. Should it do so, the collective interest of the industry in a standard confronts the private interests of an IPR holder.

Though not necessarily agreeable, these interests will in many cases be amenable, and the confrontation will not spell conflict. To simplify, the discovery of essential patents is most often settled peacefully through different licensing arrangements. The innovation system comes under strain only in cases where licenses threaten to undermine the technology's marketability. This can occur either where the terms of individual rights-holders exceed what the SDO's understanding of 'reasonable terms' or where cumulative royalty costs of several rights-holders do so. The GSM-case will provide a good case to illustrate. We will see among other things how cumulative costs of licenses were considered to threaten the extensive standardization of this mobile system. In this case, there was more than one intellectual property holder involved each with more than one patent. Is it a unique case? Tendencies in the ICT markets suggest not and experience seems to bear this out. (see below)

The conflict can be far more damaging if the IPR-holder for some reason refuses to license his technology to those wishing to utilize the standard. It should be said that this risk is somewhat more remote because a property-rights holder generally benefits when his rights are implicated by a standard. It is nonetheless a factor, and in the case of GSM, which was a *mandatory standard*, it was a particularly unwelcome factor, because a conflict between legal objectives became involved to which there was no obvious legal recourse available. We will see why the potential for this sort of conflict is improbable but how its theoretic existence can detrimentally influence the performance of IPRs and SDOs in complementarily promoting innovation.

Conceptual Framework

Before investigating the GSM controversy, however, it is important first to establish a framework in which to present and conceptualize confrontation between IPRs and SDOs. We will rely on two related areas of innovation theory here, the systems-oriented literature of innovation theory and economic theory concerned with institutions and their evolution. This discussion will however be limited to the present ends, (i) of establishing a integrated linkage between 'institutions'[1] and the ICT industry; (ii) of demonstrating that such institutions 'co-evolve' together with the changing industry and (iii) of making the claim that this co-evolution has brought the two institutions of SDOs and IPR-regimes onto a collision course in which the 'essential IPR' problem risks becoming more common.

The 'systems-theory' we refer to emerges from the extensive reappraisal of the nature and importance of technical innovation with regards to economic performance. In general, the approach argues for the conceptualization of technical and scientific change in terms of crosshatching systems. These approaches focus on different levels of aggregation and at different angles. Their common feature is that they tend to conceptualize an economy's innovation performance in terms of, "how formal institutions (firms, research institutes, universities etc) interact with each other as elements of a collective system of knowledge creation and use, and on their interplay with social institutions (e.g. legal frameworks)." (Smith, 1994) Versions of the policy oriented 'systems' literature include National Systems of Innovation (cf. Lundvall, 1992; Nelson, 1993) and Knowledge-Systems. (Foray and David, 1995) In this dynamic perspective, the position of 'social institutions' like standard development organizations or IPR regimes is explicitly recognized as playing an economic role in the 'creation and use' of new knowledge, or in other words, the development and diffusion of innovation.

This perspective is closely related to evolutionary economics and its older cousin institutional economics. These more heterodox branches of economics emphasize respectively the role of institutions and the role of technology in the economy and generally describe innovation in terms of evolutionary processes. (e.g. Dosi et al, 1988) The salient observation in terms of the present discussion is that the institutions we are concerned with not only influence the evolution of technologies and, by implication industries, but are themselves influenced to change. In other words, industry and the institutional framework or 'technology infrastructure' are seen to 'co-evolve' along with the industry.

ISSUES AND CONTROVERSIES

This section applies the idea of 'co-evolution' to the telecoms industry in order to identify the transition from the 'monopoly-provider paradigm' to the 'reregulation paradigm'. It is at this historic juncture that the IPR conflict first emerges. We will also discuss the respective roles SDOs and IPRs play as, in theory, complementary non-market mechanisms.

The Dynamic Inter-Relationship and Paradigm Shift

The first observation is that the chaos of the changing ICT markets can be construed in terms of the evolution of a 'very large technological system' in the sense of the technology- historian, Thomas Hughes. (Hughes, 1989) In this sense, technology evolves as a complex interplay of changing 'technical artifacts' (e.g. telephones, switching equipment etc), changing 'organization aspects' (e.g. away from state-controlled, monopoly PTTs) and changing 'legislative artifacts'. (e.g. competition regulation, IPR incentive structures)

In terms of the evolution of the telecom industry, a defining feature is that these aspects have been historically coordinated to a unique degree at the national level on the basis of arguments about 'natural monopoly'. Focusing on Europe, the history of this industry can be roughly divided for simplicity into two periods, before and after the publication of 1987 Green Paper (Com 290 final). The two periods can be called the "Monopoly-Providers' Paradigm" and the "re-regulation paradigm".

The legacy of the industry is intimately tied to the 'Monopoly-Providers' Paradigm, a paradigm which by all accounts demonstrated a unique degree of public-sector control. This control shot through all levels of 'artifacts' nationally, while internationally it was typified by collaboration. The hallmark of this control-collaboration paradigm is the state-controlled monopoly PTT (center of control) and their standardization organization, CEPT (center of collaboration). In this paradigm, the market was characterized by a monopoly-monopsony relationships with the PTTs as the only buyers and the 'national champion' as the only developers and suppliers of communication technologies. This market situation, which was instrumental to the IPR question, was in turn closely related to the paradigm's legislative setup, typified by strong national regulations combined with cooperative international treaty agreements. The technical factors associated with this paradigm is the single-service, plain old telephony or POT service.

The transition to the 'reregulation paradigm' is necessarily vague and certainly has not (yet) happened overnight. Nonetheless, there is a general transition, generally associated with the 'digital revolution', that can be seen qualitatively as a breaking up of the coordinating power of the older paradigm. Out of the breakup, the interplay between different elements of the evolution has become exceptionally dynamic and the evolution as a whole has been marked by discoordination (eg. new alliances, fumbling for policies, HDTV). In particular, three integral dimensions of the (r)evolution are pulling at each other:

- Changing technical conditions; these are exemplified by rapid improvements of microelectronics (see Moore's law), the multiplication of transmission modes (cellular technology), and the proliferation of new technologies and services in train with the drive towards end-to-end digitization;
- Changing market-regulatory conditions: the convergence of voice, broadcasting and data industry and an pervasive drive towards globalization. In a European context, these are marked by the otherwise independent drive towards the Single European Market (hamonization, liberalization) and the concomitant change in standardization policy;
- Changing institutional conditions; a diversity of alliances, mergers and R&D

joint ventures in previously separate sectors belie a fundamental shift in the User-Producer relationship. This is reflected especially in institutional standardization, where there is a multiplication of types of SDOs and range of actors that are involved. CEPT becomes ETSI.

Table 1 illustrates the basic shift between the two paradigms and indicates that

	Regulatory status	Institutional set-up	Market conditions	Relation to IPRs
Monopoly-Provider Paradigm	POTs Oriented	CEPT	Monopoly-Monopsony markets	IPR conflicts not a problem
	Strong regulation nationally, cooperation internationally	PTT	National champions-national PTT	"Have-made rights"
		Closed, consensus		
Re-regulation Paradigm	"Deregulation"	ETSI	Transition to duopoly and opening up the supplier markets	IPR conflict
	Cross-border harmonization	Heterogeneous participants	International actors more active	ETSI IP Regulation
	'level playing field'			

Table 1. The transition from Monopoly-Providers' Paradigm to a Re-regulation Paradigm

the IPR problem arises in the transition from the controlled atmosphere of the first to the more chaotic environment of the second. In the next section, we will discuss the respective roles that standards development organizations and intellectual property rights play in terms of ICTs and indicate that these are indeed changing.

Variety and Selection: The Roles of IPRs and SDOs

In this framework, institutions such as industry SDOs or IPR regimes are, at different levels, "institutional devices for allocation problems" in the innovation process. On the one hand, industrial IPRs are generally seen as 'appropriability mechanisms' to encourage R&D investment while diffusing information about the invention.[2] On the other, cooperative SDOs address the market's difficulty in effectively selecting new network technologies.[3] At issue is how the interplay between the legal-framework and standardization institutions becomes the object of dispute as the system evolves and how, in turn, the dispute affects the continued creation and use of 'economically valuable knowledge' in the system. (Foray, 1995)

In the case of network technologies, evolution involves on the one hand the generation of new techniques and new technological options and an ongoing reduction of these choices by selection on the other[4]. The environment in which these processes are to take place is ultimately the market. However, the market is less than optimal both in promoting variety of innovative technical ideas and in selecting out new network-based technologies. Therefore certain mechanisms have been created to address the perceived 'market failures'. To aid the market, economies have developed incentive mechanisms to help induce the generation of innovative inventions and selection mechanisms to improve the selection process. They do not necessarily install modes of coordination that prevent these processes from conflicting with one another.

The conflict between IPRs and formalized standardization is one that emerges in what Foray calls the, "encounter between the processes of technological diffusion and interdependence (which induce standardization) with incentives and organization structures (which induce diversity)." (1995) This encounter can be seen in terms of the non-market mechanisms that promote these respective processes, such that formal standardization within SDOs can be identified with the diffusion process while IPRs can be identified with the provision of diversity.

The division of labor is not perfect here, and there are certain aspects of each that increasingly affect the other process. Certain aspects of IPRs (i.e. the diffusion of technical information) align with the diffusion function while IPRs are increasingly used to promote interdependence between technologies as they lay the basis for R&D cooperation. In addition, the importance of IPRs in the ICT field is increasingly associated with this sort of use as a "poker chip" to get involved with new alliances and new markets. Formal standardization, as it moves to promote the creation of large technological systems like GSM, seems too to be moving towards an organization conduct of quasi-R&D activities.

Nonetheless, identifying IPRs and SDOs with these processes helps us to recognize the conflict between them as one involving two learning-processes at the heart of the innovation. Consider first that, especially, in the ICT, "evolution is the result of two seemingly contradictory processes: the creation of variety and its successive reduction through selection. Effective long-term adaptation requires that these two processes be kept in balance." (Carlsson & Stankiewicz, 1991) Accordingly, industrial IPRs can be identified with learning to create variety at the level of the firm and formal standardization with the market's learning to select. Conflict between these non-market mechanisms thereby risks bringing these two fundamental processes into imbalance, jeopardizing the long-term adaptation of the ICT industry.

The roles of IPRs and SDOs in ICT

In the turbulent evolution of ICT, companies are looking for market shares both in terms of access to the new technology markets and/or access to opening geographic markets. The rapidity and diversification of technical change is accompanied by greater R&D intensity, both in one's area of expertise and along expanding borders, which puts a premium on IPRs. Success in this environment is therefore predicated on strength of one's technical specialization but also the creation of alliances in neighboring technologies and in foreign markets. R&D

becomes more expensive, which emphasizes the importance of IPRs. However, so does the prospects of failure, which in turn emphasizes the success of common standards.

IPRs: Increasing significance in ICTs

IPRs cover individual R&D activity and can therefore be identified with the firm-level learning process and the production of technological variety at the aggregate level. In the changing environment of the ICT-industry, the importance of IPRs has increased and their use seems to be changing. Patenting has become at once much more strategic in order to address increasing competition as well as much more oriented around cooperative R&D, because IPRs can facilitate collaboration. It is claimed that in the increasingly global ICT markets, "IPRs are the most valuable asset an ICT manufacturer has." (Miselbach & Nicholson, 1994) There are several characteristics about the ICT field that justify the attachment of such great importance to industrial rights (especially, patents, also mask works, software copyrights).

1. Cost-structure of manufacturers involves high investment risks. The ICT field combines high R&D costs, regularly in excess of 10% of turnover, with low variable costs. One is therefore vulnerable to copying.
2. Great uncertainty about the industry's future development combined with the strong network externalities of many ICT means that not holding rights in the technology that becomes the market standard may exclude you from new markets
3. The pace of technical change means that it becomes easier to innovate around existent patents.
4. There is also increasing competition at home and abroad as markets re-regulate and become more global. It is therefore increasingly important for manufacturers in this environment of increasing competition to protect their R&D investments against misuse in other markets
5. Moreover, there is increasing combination of cooperation and competition, meaning that one wants more control over one's knowledge assets when entering increasingly common R&D joint-ventures (which are increasingly embraced to tackle high R&D costs, and uncertainty about market developments).

Formal Standardization and Selection

As with IPRs, the importance of standards development is increasing in the changing ICT environment. As we know, formal standards-making is designed to diminish uncertainty in network-oriented industries by attracting a consensus around individual technologies. In so doing, it tries to bypass the situation in which the market either does not pick up momentum because of too much variety, or it malfunctions. In the latter case, it can mitigate the *'orphan'* problem (by producing backward compatibility) and the risk of a *bandwagon effect* that chooses the 'wrong technology'. (Besen, 1995) Because a bona fide standard acts to diminished uncertainty, it lowers transaction costs for manufacturers and other actors, increases the efficiency of information flows to the user and improves productivity diminishing inefficiency associated with protracted trial and error processes in manufacturing.

(David, 1995) Standards development therefore contributes to the collective learning process, by 'reducing nonstrategic sources of variety in design', by minimizing information costs in the industry, by providing a 'focal point' for the industry among an increasing variety of solutions, and by providing legal framework for the industry to sketch long-term strategies. These are important roles in defining the 'selection environment' (e.g. Nelson, 1995) of the ICTs.

Standards development organizations however do not necessarily fulfill these roles unerringly. Standards-making has, for example, exhibited at the national level a notable industrial political function. Standardization has been used to promote home-industry and also to influence trade-patterns. (David, 1989). Where, "patents are a means by which entrepreneurs try to augment the monopoly profits from innovation by making it more difficult for potential competitors to copy or imitate." (Dosi, Pavitt, Soete, 1990) The "manipulation of standards" have been used to serve the same cause for a nation or region. The sponsorship of a successful standard can give the sponsors an advantage in their own markets while keeping potential competitors from outside at a disadvantage. At the same time, if that standard can be exported to other markets, the advantage of the standard's sponsors is also extended. This element is evident in the GSM case, which has kept especially Japanese manufacturers from opening European markets while it has spread the influence of EU based interests to other countries. (Cattaneo, 1995)

Standardization: increasing significance in ICT

Formal standardization is undergoing comprehensive changes that are most closely associated with technological convergence and the re-regulation of telecoms. Focusing on telecoms, SDOs have gone from being a monolithic feature of the monopoly-provider paradigm in which one country meant one PTT (though some things die hard) to a re-regulation paradigm which is increasingly more heterogeneous. In this situation, conventional standardization has brought more and different players into the process (cf. ETSI, ITU-T) while at the same time a new range of individual industrial interests have organized to (pre)standardize in front of the market. In this context, formal standardization is routinely seen as the bottleneck or "reverse-salient". (Bijker, Hughes, Pinch, 1987)

Increasing Scope for IPR-Problem

In this section, we try to combine the discussion from the previous two sections to claim that the co-evolution of the functions of SDOs and IPRs is precipitating the conflict between them. This discussion will preface the final section in which the GSM case will be described.

In ICTs, the structural tension between IPRs and institutional standardization indeed appears to be exacerbated by those evolutionary pressures that are reshaping the industry. Conditions such as the accelerated pace of innovation, its increasing complexity, the changing shape of markets, the changing ranks and positions of actors and, not least, changing geopolitical conditions, all contribute to making the problem both more probable and more damaging for technical innovation.

The traditional approach to the IPR-Question

Although the potential for conflict is implicit to the voluntary standards-writing process as it functions today, the risk has largely been written off as academic. Indeed, there is no record that voluntary SDOs, which have existed in different industries for about a century, had before the GSM case experienced a unsolvable conflict with an IPR holder. There are several reasons for this in the ICT industries. The main set of reasons is that, in the case of the monopoly-providers paradigm, the way telecom standardization was historically conducted simply did not raise the problem. There are three sets of considerations here. The first is that the national standards typical of Europe were a matter of "have-made rights" whereby the PTT developed technical parameters which it gave to its "national champion" to fill. It fell upon the equipment supplier to clear any IPRs it ran across developing that technology. A second consideration involves international SDOs where standardization originally was a matter of codifying technologies that had become de facto standards and providing for interoperability between different national systems at relevant interfaces. In this case, standards were based deliberately on proprietary material (cf. Cargill, 1989). However, more recently, institutional standardization has begun working in front of the market (cf. David & Monroe, 1994.) In this situation, the probability of a standard implicating an IPR has been rated at about 2 % (ETSI document). In these cases, the SDO automatically attempts to circumvent the proprietary material. If the problem is not disarmed the standard is abandoned.

A second set of reasons why the intellectual property rights problem is not necessarily 'essentially' problematic is that IPR holders have powerful market incentives to license if the occasion arises. The reason SDOs like ISO or ITU-T immediately try to reengineer the standard to avoid proprietary material, is that incorporating such material in a wide-ranging standard gives the controller of the rights a huge market advantage. Even without royalties, this situation is lucrative because it means that the IPR holder has a head start in terms of learning in a ready-made market. With licenses, the IPR holder's advantage is effectively the double of the royalty rates he receives from competitors. A second consideration is that industry standards are typically "voluntary", meaning that, should an IPR holder refuse to license on agreeable terms, the market could choose a rival standard. In short, the holder would have no obvious leverage and anything interpreted as predatory licensing would leave the holder not only alone but begrudged, a dangerous situation in a field characterized by extensive collaboration.

The compounded risk in ETSI

In this environment, SDOs no longer see the IPR problem as an academic problem. However, the fact that the touchstone for controversy emerged for the first time in the EU as the standardization organ of the Monopoly-Providers' Paradigm, CEPT, was opened up to the more heterogeneous ETSI is not altogether a matter of chance.[5] Specifically, there were at least three aspects that were related to the EU, the transition to ETSI or to the GSM technology that sparked the first controversy which we will survey in our last section. The first bears specifically on GSM, where some argue certain market aspects made it more lucrative for the IPR holder to be aggressive in its licensing approach. This observation has to do with the high stakes involved in these standards.

The second aspect bears more generally, and thus more seriously in the long term, on ETSI. ETSI is one of the three SDOs to which the EU Commission can make direct reference in its pursuit of a single market. Though it is extraordinary, ETSI does produce technical specifications—so-called 'mandatory' and 'harmonized standards'—which acquire the force of law for example via EU "Common Technical Regulations" (CTRs). This situation entails that the only access to the market is through reference to that standard. In it, the IPR-holder finds himself with increased leverage vis-à-vis the market. There is of course recourse to art. 86 if the holder is found to have abusively exercised his IPR. However, as the relevant European policy is formative and the legal procedures expensive, complicated and protracted, this mechanism provides limited security to parties reliant on the successful launch of 'mandatory' standards. This threatens to lead to the delicate situation in which patent law would be contravened by following European secondary legislation (for example, the ONP-leased line specifications). In addition, a set of voluntary standards are also directly promoted by regulation (e.g. Council Decision 93/38/CEE on public procurement), making these more or less mandatory. This means that inter alia, the market is not free to choose another standard if one is blocked by an IPR.

A third aspect is that European manufacturers are no longer guaranteed national markets in the Re-Regulation Paradigm but are dependent on ETSI standards to establish themselves in the singular European market. In terms of building the European market, it therefore becomes imperative for European players to diminish risks that these standards will be jeopardized in any way: European standards provide a technical lingua franca which allows them to grow into the wider deregulated European equipment market and secure market shares. Given the insecurity of this position, the risk that a European standard strands on an IPR is the most serious risk for those players with Europe as a home-market. But there is another aspect to their interests in this respect. These European players are also interested in establishing the European standards in the wider arena of changing world markets. Sponsoring a regional standard that later gains wider currency is beneficial both in terms of European reputation and because manufacturers can enter these markets with the advantage of a familiar technology.

THE GSM-CASE

A prime example of a case where a European standard successfully opened the door to international markets is the GSM system. The popular GSM900/DCS1800 system is based on 10,000 pages of technical specifications, covering all aspects of the mobile-system, from switching (NSS), to radio transmission and reception, to channel coding, to terminal specifications, to service recommendations etc. Further, they encompass two phases of the system's deployment and include the specifications for a related PCS network at 1800MHz (DCS1800; 'delta specs').

The popular GSM-system that those standards more or less 'blueprinted', hides an interesting history of institutional systems-building. Tracing back to 1979, one of its interesting aspects was its degree of orchestration of regulatory, political, organizational aspects that were of the same complex quality and had the same sort of committee-origin as the technical aspects integrated into the specs. Relatedly, this orchestration of committees, taken together, ran a rather serious risk on the IPR front. A decade into its history, the building of the systems found that patented technical ideas had been built into the complicated technical aspects despite certain precautions. In fact, all the technical sections of this mobile system, with the possible exception of switching, implicated IPRs!

In this situation, any manufacturer who produced, and any service provider who utilized equipment based on these standards would have been violating the IPR holder(s)'s rights if they did not get their clearance. These users would thus become liable under patent law. Considering the technical breadth of the standards and the rising intensity of R&D activity in mobile-communications, this emergence of implicated 'essential IPRs' could not have been wholly unexpected. Still, the number of IPRs the standards risked implicating and the variety of interests represented by the potential IPR holders made it clear: the risk was not academic.

Given this situation, the relevant procedures of other SDOs would have dictated that the standards be resubmitted the Working Parties to circumvent the implied IPR(s). However, such an approach was clearly impractical in this situation and for the system builders, unacceptable. Following this procedure would have entailed costly and time-consuming searches to find out which patents could claim to be 'essential' to the standards (cf. IPR searches). Discovering this would launch an even more laborious process of finding out whether or not it would be possible to reengineer the standards, which were still under development, to avoid the 'essential IPRs', perhaps only to implicate others. This would have been inexorbitantly costly and would have hopelessly delayed the awaited standards even if the IPRs could have been circumvented. The alternative approach would be to secure licenses for the IPRs. The story behind the system builders' strategy in securing such licenses and the strategies of the IPR holders forms an instructive picture of the forces involved in the IPR conflict.

A great deal was riding on the GSM standards. In turn, these standards were riding on the availability of IPRs. This situation indicates in general the increasingly important status of standards in the changing market. In the confines of the particular case, it was a situation that did not pass by the IPR holders unnoticed. IPR holders interested in profit maximization on R&D investment found themselves commanding increased leverage in licensing arrangements in the ETSI environment. This situation placed the PTTs, who had originally backed the standards and who had substantially shaped them before other parties were allowed in 1987, in a highly unaccustomed and highly vulnerable position. Fears and rumors ran wild among the PTTs, who were more accustomed to passing on the IPR risks in contracts to their suppliers (see 'have made rights') when one IPR holder started demanding separate undertakings during the first round of bidding for GSM contracts.

Scenarios: Cumulative Costs and Refusal to License

In this case, the adoption of the GSM standards faced two basic sets of risks in terms of 'blocking IPRs'. In the first scenario, the greater risk was that the cumulative cost of royalties for all implicated IPRs could raise the cost of GSM equipment beyond that which the market would bear. In the event, this appears to have been a well founded worry as the number of IPRs claiming to be "essential" was very large and was spread between several IPR holders: the North American radio-communication manufacturer and the single largest IPR holder, Motorola, alone claimed "two dozen to 30" such patents.

This raises a second set of scenarios. Here, the fear is that a single IPR holder might exercise its right to refuse to license its proprietary technology at all or might refuse to license at fair rates, with reasonable conditions and in a nondiscriminatory way. As shown above, having one's technology incorporated into a wide ranging standard in one fell-swoop gives the IPR holder a tremendous advantage over potential competitors. Yet, despite such a windfall, there are still plausible though improbable reasons for an IPR holder to refuse to license. It has been indicated that the size of the market (1 billion pounds for the UK alone) invited IPR holders to "change their normal approach to licensing." (Buttrick, 1993)

Several generic scenarios in which IPR holders might choose to decline to license can be considered. One scenario is that the IPRs implied are so central to the firm that being forced to license these 'crown jewels' would be tantamount to selling out the company base. Another possibility, is that, in addition to the patent(s) contravened IPRs, a firm also holds rights to a rival type of technology that is more central to the firm. If the firm reckons that a standard which is gaining support in the SDO would compete with its other technology, it might have active interests in suppressing this potentially rival standard. A last scenario involves out-and-out "warfare" between geographic standardizing blocks. Put bluntly, the hypothetical situation could emerge in which one firm, who controls patents in Europe as well as say the US, chooses to support the development of standard in the US while refusing to license a technology that was essential to a rival Euro-standard. By refusing to license, it might hope that the EU would abandon its own standardization efforts in favor of the rival standard.

However, the likelihood of these and other conceivable scenarios actually occurring are very low for a variety of reasons. The most important of these is that such a firm would seriously upset the 'positive sum' nature of standardization and risk a sort of ostracism in the relevant market. The threat of informal reprisal by its competitor of such an IPR holder can be serious, as will be illustrated in the case of IPR holders involved in the GSM standards. Whatever the practical likelihood of such scenarios, the point remains that these situations are theoretically possible and that, should they emerge, there is no immediate legal recourse to force the IPR holder's compliance. The potential is there to take the IPR holder before a compe-tition court in the EU for 'abusive exercise of a dominant position', but such an action is highly contingent on other issues and, besides, is resource intensive. But moreover, there is no clear precedence in the court to which to anchor such an action.[6]

Collision Course

The GSM-standards involved a dedicated, comprehensive systems-building project in which technical, political and institutional aspects of the system were engineered in great detail.[7] In our context, there are three immediate aspects about the process that visibly raised exposure to the IPR problem. The first is that the broad scope of the standards not only opened for the implication of a large spread of individual IPRs, it also risked implicating so-called "system patents" which cover concepts that pervade the system such as those claimed for TDMA (Time Division Multiple Access).

A second aspect is the level of complicated technical detail with which the 280 recommendations involved are drawn up. The greater the detail of the standards, the more they resemble the formulation of patents and thus the greater the scope for the two to come into conflict. This 'over-specification' was to a degree deliberate, owing to the overriding concern that the system must connect with existing analog networks as well as proactively to ISDN networks. Not only must the GSM system connect; moreover, it must be interoperable despite the idiosyncrasies built into the dissimilar national networks. As an indicator of how formidable such a task is, the GSM system as laid out in the 10,000 pages still needs to be tweaked in order to interoperate properly.

A third, related aspect suggested by the timeline (cf. Annex), is the length of time that standardization took. Duration of standardization increasingly becomes a factor because, while standardization is under way, active patent-directed research continues and perhaps is even augmented by the knowledge that a standard in such and such a field is under development; as noted, holding an essential patent is lucrative. This is risky in today's climate of proliferate technologic development because, while a standard needs to keep on top of this development, by doing so it risks duplicating patents that are being created outside it.

The IPR Minefield

The wide breadth, the detail and the duration of the GSM standards raised the likelihood of collision with IPRs. On the other side of the equation, there are several aspects about intellectual property rights that simultaneously are raising the risk of a clash. The first has been mentioned. Technological change in the ICT field has become self-reinforcing and the tempo of change is dramatic. This activity is stimulating the production of patents. Patenting has been active especially in an area like mobile communications, where the majority of technical principles are, according to the industry, 'well understood' but where new technical alternatives are appearing that open market prospects.

The composite picture is one of an increasingly densely-packed 'IPR Minefield' through which a standard must navigate. While R&D intensifies and the numbers of IPRs increase, patents can live for up to 20 years. This means an augmented reserve of 'active' patents which might block standards. The inclination to try to capitalize on these monopoly rights is also rising. One factor that makes a holder more inclined to exercise his rights in this forum is the uncertainty of the markets. No longer assured a piece of the market, IPRs take on increased importance because they give their holders access to markets made turbulent by shortening product horizons, technical convergence, different degrees of liberalization and general regulatory uncertainty. In this situation, if a manufacturer does not produce at least one 'major' innovation every six months (as opposed to one every four years 5 years ago), that firm is pushed out of the game. In a manufacturer's metaphor, IPRs function as 'poker chips' that allow you to enter a market and vie for market shares.

Thus, features connected both to standardization and to the exercise of IPRs, and many more beside, contributed to the likelihood of collision. In the GSM case, its systems builders tried to manage this risk as best they could. This aversion to the IPR risk is first seen in 1986 in which they deliberately chose a "broad avenue" technical approach in order to avoid IPR questions that would have resulted had they supported one of the 8 proprietary prototypes then demonstrated. The second, was a common tact on the question of IPRs which was launched in the first round of tendering for GSM contracts. We look closer at the controversy.

Towards Dispute

The IPR dispute began to materialize during the first commercial contacts between buyer (Telecom operators)-supplier (equipment suppliers) for the provision of equipment based on the GSM standards.[8] It was in these bidding rounds that the number of IPRs impacted by the GSM standard began to emerge and where the systems-builders became worried.

What provoked the confrontation were the terms governing bidders' freedom to exercise their IPRs that were employed unilaterally by the 17 participating PNOs. These terms were grounded in an agreement in which the PNOs (at the time, the PTT administrations) of 15 CEPT countries entered in 1987, directly before the hand over to ETSI. This so-called GSM Memorandum of Understanding (MoU) supplanted an earlier four party agreement from 1985 and put into place the logistics of a coordinated launch from the PNOs point of view. In it the signatories committed themselves to a common organizational line on the deployment of the GSM system.

The MoU was designed by the PTTs to minimize coordination risks associated with the orchestrated launch of a complex technical system. It was imperative to the success of the GSM system that the launch be synchronized, that equipment-type be proven compatible and that there was a rolling commitment to its future development of system. Another important area of risk was IPRs[9]; approaches to this area were to be harmonized among themselves, the buyers.

The MoU functioned as an agreement among purchasers in which 100% of the market was represented. It was in other words a monopsony arrangement. Together these countries' PNOs commanded a position analogous to that held by individual PNOs under the monopoly provider's paradigm, i.e., before the effect of the 1987 Green Paper began to be felt. Instead of one PNO codifying technical specifications and passing these as "have made rights" to their national champion as would have been the case during the dying paradigm, the initial 17 signatories of the MoU in effect extended this practice to the international arena. Based on the MoU, all potential equipment suppliers were collectively presented with similar terms in the lucrative contracts for the provision of GSM equipment. It is clear that these terms were meant to defuse the risk that IPRs could pose to the collective launch of this "over specified system" and further it is clear that these terms had their roots in the traditional producer-customer relationship alluded to above. But despite this degree of conventionality, the terms pertaining to the equipment suppliers' exercise of IPRs that were codified in these preliminary contracts proved contentious for some of the manufacturers. This contention in turn led to the unconventional situation in which the constellation of buyers was challenged. What was contentious for the IPR-holders was that the contracts specified that equipment suppliers were obligated to undertake to license any "essential" patents royalty-free within the CEPT area and to license to all-comers outside the CEPT area at "fair, reasonable and on-discriminatory terms."

For some "national champions", such terms were familiar from the monopsony-monopoly "have-made rights" system that the Green paper and ETSI were avowedly in the process of building down. These players did not at first seem inclined to challenge these terms, despite the fact that they found them 'unfair' (e.g. Alcatel). The individual reactions of different suppliers must however be seen in terms of a set of factors that includes: how many patents a manufacturer held that could be construed as "essential" to the GSM standard; which technical area they were in; and, relatedly, the orientation of their IPR strategies.

Response and Recriminations

It was thus when the US-based Motorola claimed that it held "between two dozen and 30" patents that it construed as "essential", that the industry realized more fully than before that the potential for conflict between IPRs-SDOs was a reality. It should be appreciated that strong market logic drove Motorola to seek to maximize profits on its IPRs. First there was the sheer number of patents (3 times as many as the nearest rival) representing considerable R&D investment. This fact alone effectively raised Motorola's 'ante' and implied that it would want a larger part of the pot. Further, the technical area in which these patents were concentrated, was important. The reason for this has to do with the different types of payoff structures connected to different sorts of technologies. Motorola points out in this context that it is not involved in switching-plant technology, where a company can count not only on the initial technological solution they ship but further on the long term revenues derived from proactive software enlargement contracts. In this scenario, the importance of deriving returns on the IPRs of the initial technology arrangement is not decisive owing to the fact that returns are secured over longer periods. Instead, Motorola contends that returns to R&D investment in its main area of radio terminal equipment are more oriented on a one-off basis. Here the claim is that you derive revenue only on the equipment you sell. In light of its calculations of this market, Motorola was dissatisfied with market prospects and was therefore unwilling to forfeit the additional returns afforded by licensing royalties.

This raises a third characteristic about Motorola, which contributed it to becoming the most vocal opponent to the GSM-MoU signatories' terms, namely Motorola's aggressive in-house IPR policy coupled with its lack of market shares in Europe. In Motorola, patenting is an expressed prime objective of the R&D culture, as in this milieu managers are measured according to the number of patents filed. Among European corporations the practice had been quite the reverse. In fact, " until very recently, patenting has deliberately been avoided, " (Deiaco, 1994) by European manufacturers, largely due to their respective positions in Balkanized telecom markets governed by a 'control-collaboration' logic. In addition to the diminished incentive to patent built into the market-structure, patenting has also been deliberately avoided in order to minimize transfer problems to their buyers, the PNOs, who have been interested in adapting the technology they contract to their needs.

It was against the relatively inactive patenting culture in Europe that Motorola's 'aggressive use of patents' clashed in the GSM case, and it can be said that as a result, "Motorola's behavior triggered a new era of awareness of IPR(s) in telecommunications both in Sweden and abroad." (Deiaco, 1994) Motorola's behavior in this sense should be seen in terms of its position as both somewhat of a pioneer in radio-signal technologies as well as somewhat of an outsider to the European market. Given this situation and the active use it traditionally makes of patenting, it is not surprising that this North American company decided to utilize its patents to gain access to market shares in the dawning European market. Nor is it surprising that, given the ongoing changing market situation especially in Europe, such IPR strategy has become actively adopted by all 'players' as they vie for market shares.

The adoption of the GSM standards therefore represents something of a watershed in the way European companies generally approach intellectual property rights. Partially because of this watershed element of the situation, Motorola's challenge to the IPR-terms on offer gave rise to considerable dispute. A major factor however in this dispute was the underlying division between companies with Europe as their home market and those, like Motorola, that had their home-markets elsewhere. In the common parlance, it was one front of the "warfare" over technology that is underway beneath the surface of the relationship between the US and the EU.

It became apparent that the emerging GSM system was extraordinarily exposed to the risk that either "cumulative" licensing costs would price GSM out of the market or that IPRs would not be licensed. The system was indeed perceived to be extremely vulnerable. Therefore when Motorola refused the terms of the MoU undertaking, demanding separate undertakings for individual contracts, the systems builders' concerns reached a pitch. A series of accusations and recriminations was sparked between 1988-89, that called into question Motorola's strategy and the fate of the GSM system. Was this US based manufacturer refusing to license its patents? Would the terms it demanded give it such an advantage that it would take de facto control of the European market? Against a background where the system was more vulnerable than expected, rumors emerged that Motorola was trying to use its "essential" intellectual property rights in order to hold the GSM standards process to ransom.

The position of Motorola, with the tacit support of some of the suppliers, came the claim that the PNOs were abusing their position in CEPT as a monopsony to dictate licensing conditions, and that the rumors against it were in part ignorant and in part libelous. It is difficult to evaluate either the substance of the anti-Motorola rumors or the licensing terms Motorola was demanding. What should be noticed is that the general consensus emerged that the MoU was a "failure". Motorola et al. pursued their own licensing procedures and the price of the system has far from rising beyond that which the market could bear because of royalty costs, has proven very competitive and has spread to some 119 countries. This is to say that the cumulative price of royalties inside EU did not prove prohibitively high, and that licensing practices did not stop the GSM system from being propagated to other countries including parts of the US market. The lasting result of this dramatic demonstration is that the IPR conflict can emerge with increasing frequency and with increasing scope for fundamentally influencing the direction of certain types of technological development.

Beyond GSM

The IPR problem therefore did not pose a insurmountable obstacle to the diffusion of the GSM system. It did however lead directly to a volatile debate involving the development of ETSI's IPR Policy. This controversy, which led to a diplomatic situation involving the US and the EU, demonstrated the importance of this issue. But this is another story (see, Iversen 1999).

In addition, the risk of the IPR question has since emerged both in formal, industry level and bilateral efforts at creating an industry standard. Also a part of the GSM standards, a dispute arose concerning not patents but software copyrights for a voice-modulation technology. The DVSI case involved a small company and the degree to which it had to reveal all aspects of its software code (its 'crown jewel') as a condition for using its technology in the final standard. This case is significant in at least two respects: first, it seems to be the first case involving IPRs and standardization that went to court and second it signalled that not only patents but other industrial rights could be involved in this sort of dispute.

The IPR problem however does not only emerge in ETSI. During ISO's standardization of MPEGII audio-standards, a situation is said to have emerged involving MPEGII standards connected with film/televisions, for which ca. 25 actors claimed to have nearly 100 patents that were potentially 'blocking' for the standards. It was agreed that the question would be considered ex post the launch of this standard, and it is not known what has happened here.

Finally, it is important to realize that the risk is not isolated to international standardization. In 1995, a case is said to have emerged in the American industrial standards association, Video-Electronics Standards Association (VESA). This conflict involved the Federal Trade Commission vs. Dell computers (Decision C-3658). It revolved around the latter's usage of its patents after its participation in the standardization of the VL-bus technology that links the microprocessor to peripherals in 486 PCs. In this landmark case, Dell was forced to drop its claims, by which it had sought to extract royalties from its rivals after the standard was finished. This case illustrates that the conflict can also arise in industry standards associations, which are flourishing in today's ICT climate.

These cases illustrate that the tense relationship between Intellectual Property Rights and standardization involves both different types of IPRs (patents and software) and different levels of standardization: regional (ETSI), international (ISO) and industrial level (VESA). This indicates that the IPR problem signaled by the GSM standards, indeed is gaining scope for controversy. This is not a problem for which there is any easy solution. The recommendation made here is that the problem be studied in terms of its relation to the wider innovation process.

CONCLUSIONS

This chapter has investigated some dimensions of the IPR problem faced by SDOs within the context of the changing ICT field. The chapter indicates that SDOs and IPR-regimes are becoming increasing important components of the evolving ICT industry. It was demonstrated that coevolution brings IPRs as a mechanism associated with the creation of technical variety, increasingly into conflict with SDOs, which are associated with the delicate 'selection environment' of network technologies. This potential for conflict was sketched with reference to the GSM-case. Finally, several other cases of conflicts were noted to support the central thesis that the scope for conflict is increasing and spreading to other fora of standardization.

The IPR-problem indicates an increasing discoordination between two central aspects of the 'technology infrastructure'. The impression that the co-evolution of IPRs and SDOs is bringing their functions into conflict is important and should be studied further. Factors that should be taken into consideration include: (i) increasing regulatory activity at the global level; (ii) the changing makeup among participants in the SDOs, especially the importance of their market-structures; (iii) the conflict in terms of 'organized markets' (Lundvall, 1993); (iv) the increasing pre-standardization consortia or fora and SDOs, like ETSI; (v) the importance of evolving IPR-strategies in the ICT-industry.

Annex: Historical Context of the GSM-Standards

Highlights

1946: First civilian mobile system launched in 1946
1979: 900 MHZ BAND RESERVED BY WARC
       AMPS launched (BELL LABS)
1981: NMT (coop. Between Scandinavian PTTs and some manufacturers)
1982: FIRST MEETING OF Group Spécial Mobile (GSM) IN STOCKHOLM
1985: TACS (AMPS based)
1987: GSM OPTED FOR 'THE BROAD-AVENUE' DIGITAL APPROACH
       GSM MoU
1988: ETSI INSTITUTED
       IPR CONFLICT COMMENCES WITH REFUSAL OF MoU TERMS
1989: GSM TRANSFERRED TO ETSI
1992: GSM PHASE I STANDARDS (FINAL)
1993: GSM PHASE II STANDARDS
       STANDARDS COMMITTEE ESTABLISHED FOR UMPTS
1996 FINAL TESTING OF GSM

ENDNOTES

1. By 'institution', we mean formal structures such as patent regimes or standards development organizations. Cf. North, 1990.
2. The 'market-failure thesis', in which Neoclassical economics pioneered the idea that the market tends by nature to under-supply R&D activity. Lacking an incentive to invent, the exploration of different technical paths and their relative worth would not be carried out to generate a sufficient degree of the variety. See the seminal works of R. Nelson (1959) and K. Arrow (1962). A more current approach focuses on IPRs as creating a "market for

knowledge" (Gerowski, 1995). In the context of ICTs, it is increasingly relevant to focus on IPRs as a tool to facilitate R&D collaboration.

3. For a discussion of "some new standards for the economics of standardization in the information age", cf. David & Greenstein (1990) ; for a discussion of difficulties of public policy in selecting technologies for the market, see David's discussion of "narrow windows, blind giants and angry orphans" referred to in (David, 1987).

4. For a discussion of "Network Externalities, Competition and Compatibility", see Katz and Shapiro (1985). For a discussion of "high technology and the economics of standardization", see Cowan (1992).

5. The reader is directed to Besen, 1990 for an early discussion of the transition to ETSI.

6. To make up for this lack, the EU published a Communication (Com 455. 92) that provided a set of general guidelines indicating the basic rights and responsibilities for SDOs and IPR-holders.

7. See the Annex, for a timeline of the key phases of the GSM-standards.

8. The Telecom Operators means at this time the Public Network Operators (PNOs) who had initiated bidding among potential suppliers on the preliminary GSM equipment in 1988, signaling that the main dimensions of the system had already been decided. Incidentally, this began to occur at the same time that manufacturers were first being co-opted into the standardization process in the same transitionary momentum that gave rise to ETSI.

9. Article 9 of the MoU, " Co-ordination of IPR policies".

REFERENCES

Amory, B. (1992). Telecommunications in the European Communities: the New Regulatory Framework. *International Computer Law Advisor*, 1991-2: 4-16.

Arrow, K. (1962) Economic welfare and allocation of resources for invention. *The rate and direction of inventive activity*, Princeton University Press. 609 – 25.

Besen, S. M (1990). The European Telecommunications Standards Institute: A preliminary analysis. *Telecommunications Policy*, 14:521—30.

Besen, S. M. and Raskind, L.J. (1991). An introduction to the law and economics of intellectual property. *Journal of Economic Persepecives*, 5.1. Winter: 3.

Buttrick, R. (1993). *ETSI, Intellectual property rights and Standardisation*. General Distribution Document: British Telecom. London.

Bijker, W.E, Hughes, T. & Pinch, T (1987). *The Social Construction of Technological Systems*. (Fourth Printing). Cambridge, Ma: The MIT-Press.

Carlsson, B. & R. Stankiewicz (1991). On the nature, function and composition of technological systems. *Journal of Evolutionary Economics,* 1, 93 - 118.

Cargill, Carl F (1989). *Information Technology Standardization: Theory, Process and Oranizations*. Digital Press.

Cohendet, P., Héraud, J.A. & Zuscovitch, E.. (1992). Technological learning, economic networks and innovation appropriability, 66 -76.

Collins, H. (1987). Conflict and Cooperation in the Establishment of Telecommunications and Data Communications Standards in Europe. In Landis Gabel (ed). *Product Standardization and Competitive Strategy*. North Holland: Elsevier Science Publishers B.V.

David, P.A. & Greenstein, S. (1990). The Economics of Compatibility Standards: An Introduction to Recent Research. *Economics of Innovation and New Technology*, 1(1 & 2), Fall, 3-42.

David, P.A. & Monroe, H.K. (1994), "Standards Development Strategies Under Incomplete Information - Isn't the Battle of the Sexes Really a Revelation Game?," MERIT Research Memorandum 2/94-039.

David, P.A. (1987) Some New Standards for the Economics of Standardization in the Information Age.   In P. Dasgupta & P. Stoneman (Eds.) *Economic policy and technological performance*, Cambridge; Cambridge University Press,   206-234.

Dosi, G., Pavitt, K., & Soete, L. (1990). *The economics of technical change and international trade.* Hertfordshire, GB. Havest Wheatsheaf.

Deiaco, E. (1994) *Profit form Innovation.* (OECD Paper #25). Paris.

ETSI (1993). Intellectual property rights Undertaking. (JAC05101). Sophia-Antipolis, Fr.

ETSI (1993). Intellectual property rights Policy and Undertaking.(etsipol1). Sophia-Antipolis, Fr.

ETSI (1994). Intellectual property rights Policy (DD/296B/94/FA/msm). Sophia-Antipolis, Fr.

ETSI (1994). Chairmanship's Informal Report of the Special Committee on IPR to the GA Chairman (GA20/94/FA/msm). Cannes, Fr.

ETSI (1994). Final Report of the Special Committee on IPR (GA20/94/FA/msm). Cannes, Fr.

Farrell, J. & Saloner, G. (1985). Standardization, Compatibility and Innovation. *Rand Journal of Economics, 16:1*, 70-82.

Farrell, J. (1989). Standardization and Intellectual Property. *Jurimetrics Journal, 30*, 35-50.

Good, D. (1992) How Far Should Intellectual Property Rights Have to Give Way to Standardization: the Policy Position of ETSI and the EC. *EIPR,9*, 295-7.

Good, D. (1991) 1992 and Product Standards: a Conflict with Intellectual Property Rights? *EIPR,11*,   398-403.

Geroski, P. (1995). Markets for Technology: Knowledge, Innovation and Appropriability. In P. Stoneman, (Ed.) *Handbook of the economics of innovation and technolocial change* (Blackwell: Cambridge, USA.

Greenstein, S.M. (1992). Invisible Hands and Visible Advisors: an Economic Interpretation of Standardization. *Journal of American Society for Information Science.*

Grindley, P. (1995). *Standards, strategy, and policy: cases and stories.* Oxford University Press.

Hawkins, R., R. Mansfield & J. Skea (Eds.). (1995) *Standards, innovation and competitiveness*: Edward Elgar..

Hughes, T. P. (1989) The Evolution of Large Technological Systems. In Bijker, Hughes & Pinch. (Eds) *The Social Construction of Technolgical Systems.* Mass: MIT Press. (4th pr). 1993.

Holleman, R. (1996). PatCom meets the challenge of standards patent rights issues. *IEEE Standards Board Patents Committee.* http://standards.ieee.org/reading/ieee/sB/Apr96/patcom.html

Irmer, T (1994). Shaping the Future Telecommunications: The Challenge of Global Standardization. *IEEE Communications Magazine. 32*, 20-28.

Ivesen, E.J.(1999). Standardization and Intellectual Property Rights: ETSI's controversial search for new IPR-procedures. To be given at the *1st IEEE Conference on Standardisation and Innovation in Information Technology SIIT '99*, in Aachen, Germany, September 15-17, 1999.

Katz, M..L. & Shapiro, C (1985). Newtork Externalities, Competition and Compatibility. *American Economic Review*, (75,3). 424-440.

Karjala, D.S. (1987). Copyright, Computer Software and the New Protectionism. *Jurimetrics Journal. Fall*, 33-96.

Lehr, W. (1996) Compatibility standards and industry competition: two case studies. *Economics Innovation and New Technologies, 4.*

Lehr, W. (1992). Standardization: Understanding the Process. *Journal of the American Society for Information Science. 43* (8), 550-555.

Lundvall, B.Å. (Ed.) (1992). *National Systems of Innovation: towards a theory of innovation and interactive learing*. New York. Pinter Publishers.

Lundvall, B.Å. (1993). Explaining interfirm cooperation and innovaiton: limits of the transaction-cost approach. In. G. Grabher (Ed.). *The Embedded firm: on the soicoeconomics of industrial networks*. London. Routledge.

Matutes & Regibeau. (1988). Mix and Match: Product Compatibility without network externaliteies. *Rand Journal of Economics, 221.*

Miselbach, R. & Nicholson, R. (1994) *Intellectual Property Rights and Standardization.* 2nd Update.(Published by the authors)

Mouly, M. & Pautet, M.B. (1992). *The GSM System for Mobile Communications:A Comprehensive Overview of the European Digital Cellular System*. (Published by the authors) Palaiseau, France

Nelson, R.R. (ed. 1993) *National Innovation Systems. A Comparative Analysis*. New York. Oxford University Press.

Nelson, R.R. (1995). Recent evolutionary theorizing about economic change. *Journal of Economic Literature, 33*, 48-90.

North, D.C. (1990). *Institutions, institutional change, and economic performance*. Cambridge. Cambridge U. Press.

Pelkmans, J. (1987) The New Approach to Technical Harmonization and Standardization. *Journal of Common Market Studies. XXV*, 3, 249-265.

Pogorel, G. (Ed.) (1994). *Global telecommunications strategies and technological changes*. Elsevier Science BV.

Prins, C. & Schiessl, M. (1993). The New European Telecommunications Standards Institute Policy:Conflicts between Standardization and IPRs. *EIPR. 8*, 263-266.

Scherer, J. (1992) *Telecommunication Law in Europe*. (2nd ed) Brussles: Baker & Mckenzie..

Schmidt, S.K. & Werle, R. The Development of compatibility standards in telecommunications: Conceptual framework and theoretical perspective. Mimeo.

Schmidt, S.K. & Werle, R. (1998). *Coordinating technology: Studies in the International Standardization of Telecommunications*. MIT: Cambridge, Mass.

Schmidt, S. & Werle, R. (1993). Technical Contoversy in International Standardization.MPIFG Discussion Paper. 93/5. Cologne.

Smith, K. (1994). *New directions in research and technology policy: Identifying the key issues.* (STEP Report 1/94).

Scotchmer, S. (1991). Standing on the Shoulders of Giants: Cumulative Research and the Patent Law. *Journal of Economic Persepecives: 5*.1, 29.

Solomon, R. J. (1987) New Paradigms for Future Standards. *Communications & Strategies: 2*, 51-87.

Stuurman, K. (1992) Legal Aspects of Standardization of Information Technology and Telecommunications: an Overview. *CLRS*: 1-10.

Teece, D.J. (1986)"Profiting from Technological Innovation," *Research Policy*, 15:6, 285-305.

Tuckett, Roger. (1993) ETSI's IPR Policy: The Implications for Companies using European Telecoms Standards. *Patent World, 10*, 23-7.

Tuckett, R. (1994) Standardization and IP: ETSI's IPR Policy and Undertaking in the Context of EC Law and Policy. (Working Paper).

Tuckett, R. (1992). Access to Public Standards: Interoperability Revisited. *EIPR 12*, 423-7.

Von Hippel, E. (1988) *The Sources of Innovation.* Oxford University Press.

Wallenstein, G.D. (1990). *Setting Global Telecommunications Standards: The Stakes, the Players and the Process.* Norwood Massachusetts: Artech House.

Weiss M. & Cargill, C. (1992). Consortia in the Standards Development Process. *Journal of the American Society for Information Science. 43* (8), 559-565.

OECD (ICCP series #25) *Information Technology Standards: the economic Dimension* , Paris 1991.

EC Commission (Oct. 1992) *Intellectual property rights and Standardisation.* (COM (92) 445 final).

EC Commission (1987). *Green Paper on the Developmnet of the Common Market for elecommunications Services and Equipment.* (COM (87) 290 final.)

EC Commission (1994). *Europe's Way to the Information Society: an Action Plan.*(COM (94) 47final).

EC Council Recommendation. (1987) *On the Coordiatned introduction of public pan-European cellular digitalland-based mobile communications in the Community.* (87/ 371/EEC).

EC Council Decision. (1987). *On Standardization in the field of information technology and telecommunications.* (87/95/EEC).

Guidelines on the Application of EEC Competition Rules in the Telecommunications Sector. (97/C 233/02).

Cen/cenelec. (1992) *Memorandum # 8. Standardization and intellectual property rights.* Bruxelles (ed 1).

Soete, L & Arundel, A. (eds). (1993).*SPRINT Maastricht Memorandum: an Integratred Approach to European Innovation and Technology Diffusion Policy.* CEC DGXIII: Publ EUR 15090 EN 1993.

Chapter VII

Standards, Strategy and Evaluation

Robert Moreton
University of Wolverhampton

INTRODUCTION

The purpose of this chapter is to define the context of standards within an Information and Communications Technology (ICT) strategy and to suggest how the benefits arising from the use of standards might be evaluated. Within any organization, standards will be defined at a number of different levels, dependent upon the focus/span of operation. Classically, standards might be defined at three levels:

- *strategic:* the standards that should be used for all systems across an organization, including for instance standards which apply across national boundaries;
- *tactical*: standards which might apply for systems in a more limited context, such as a regional supplier;
- *local:* standards chosen in restricted or exceptional circumstances to satisfy the needs in a specific location.

This distinction is not always clear cut, and may be applied iteratively, dependent upon the context of use. For instance, a Business Unit will define its own strategic standards, or standards to support its ICT strategy. These 'strategic standards' will, of course, be defined in the context of the organization's 'strategic standards'. The local standards will 'inherit' characteristics of the strategic standards (which may be national or international in scope). It is our contention that in order to be successfully promoted, ICT standards need to be formulated within the context of an ICT strategy. (By 'ICT strategy', we mean the use and management of ICT by an organization to achieve its desired goals in a changing and competitive operational environment.) This theme forms the main basis for the discussion within this chapter on the benefits and evaluation of ICT standards.

The purpose of this chapter is to describe the elements of an ICT strategy as they relate to ICT standards, the benefits that can be gained by defining and implementing the strategy, and the factors that have to be taken into account as decisions are made about the strategy and its attendant standards. Formulating an ICT strategy includes making decisions about which ICT functions are needed by the organization to help achieve its goals, how the organization should migrate from its existing ICT base, what standards are needed, and how the functions should be procured. It does not, however, include detailed plans for installing new items of, say, system software nor for developing new applications software. These decisions are to be made at the appropriate tactical (e.g. business unit) or local (e.g. departmental) level.

We have not included application-development priorities in our definition of an ICT strategy because we believe that, in the long term, application-development skills will be spread throughout organizations. Decisions about which applications to develop or enhance next will therefore be taken by individual managers. The main role of the ICT department will be to develop and maintain the 'ICT architecture and infrastructure' that allows these developments to occur.

Nevertheless, an agreed set of application priorities will be a necessary input to developing an ICT strategy. The priorities should be set as a result of a strategic systems planning process in which the (central) ICT department will play a significant role. The ICT component of strategic planning is well documented, e.g. Earl (1989), Peppard (1993), Robson (1997), Ward et al (1997).

THE IMPORTANCE OF AN ICT STRATEGY AS A CONTEXT FOR STANDARDS

In order to be successfully promoted, ICT standards need to be framed within the context of an ICT strategy. We believe that there are three main reasons for this:

First, users are demanding a much faster response to their needs either for new software applications or for enhancements to existing applications. In addition to providing administrative support, computer applications are now essential for the day-to-day operation of most organizations, and have a direct effect on their 'business' goals. In a commercial context, they can also be the key to achieving a competitive advantage. The growing importance of applications software means that changes in business strategy now have a far greater impact on the software needed to support the business. As a result, systems departments are expected to be able to react much more quickly to new requirements.

Second, expenditure on ICT continues to be a significant (and increasingly pervasive) element in the organizational budget.

The third reason for the growing importance of an ICT strategy is that the lack of ability for current systems to interwork is seen as a major problem that prevents organizations from making best use of their ICT investments. For instance, the increasing demand for access to data, accentuated by the rapid development of the internet and world wide web, highlights the need to make data accessible, accurate, consistent (and secure) — in other words to ensure that standards are defined for data and data management.

THE BENEFITS OF STANDARDS WITHIN AN ICT STRATEGY

An ICT strategy utilizing supportive ICT standards provides benefits in many areas, the most important of which are:

- *Business demands will be responded to faster* because a strategy helps to minimize the variety of systems and software being used, which means that the ICT department can concentrate its specialist skills onto a smaller number of areas. Hence, projects are less likely to be delayed because staff with the required skills are working on other projects. Equally, where data is perceived as a strategic asset, data standards can take account of information requirements at a strategic level, and data management can be aligned closely with the systems and information planning processes (Moreton, 1989). Sweden Post, for instance, has established organizational standards which apply across its 5 business areas operating in 30 different regions.

- *Expenditure on ICT can be reduced* by minimizing the need for different software products that perform basically the same function, by minimizing the costs of replacing (or renewing) applications, and by allowing bulk discounts to be obtained wherever possible. One multinational organization with which the author is familiar saved over $4.5 million by standardizing the software-infrastructure products used throughout all its offices worldwide. Buying the software licences centrally enabled substantial discounts to be obtained.

- *Training costs may also be reduced,* both for ICT staff and for the users of the software. Training costs usually represent a greater investment than the cost of the software itself. In particular, a lack of standardization can result in very high training costs when staff are moved from one development environment to another. Training costs in the user community can also be minimized if all applications use the same user-interface conventions. In this way, staff moving from one department to another will not have to undergo extensive retraining before they can use the applications in their new department. Another organization in the financial services sector, has recently conducted a review of its total expenditure on an analysis tool. This amounted to around $135,000 for software licences and $270,000 for hardware, but more than $840,000 for training.

- *Interworking between applications is facilitated* by an ICT strategy because the strategy will ensure that the products selected can be interlinked. Often, the full extent of the need for an application to interwork with other applications is not apparent when the application is designed. Using standard products that conform with the strategy ensures that, when future requirements for interlinking arise, an application will not require major modifications and that the need to create a bespoke interface between two applications is minimized. One organization we spoke to during the research told us that the lack of an ICT strategy in the past meant that it was now faced with the problem of supporting three different systems environments. Integrating the applications will be impossible until the ICT infrastructure has been rationalized. Interfaces to external systems may also present a problem.

One organization told us that it was in the process of rationalizing its ICT infrastructure in order to achieve these benefits, and to ensure it can develop an effective intranet. This organization operates in a business sector where deregulation led to huge growth in new demands for applications from the business managers. To meet this demand, the systems department had been forced into buying packages that did not fit into its chosen architecture. Moreover, a recent merger had further complicated the situation. With the prospect of further mergers in the near future, the systems department is anxious to have a consistent and comprehensive software architecture in place before it has to cope with the need to integrate the different systems.

- *A more flexible choice of hardware will be possible* with a defined ICT strategy. Few large organizations have a single vendor policy for computer hardware. Even those that use one vendor's equipment for their mainstream computing find that they need other suppliers for scientific, technical, or manufacturing applications. Others, such as public service organizations, have a business policy to reduce their reliance on a single hardware vendor. To run the same software on different suppliers' hardware (or even on different-sized hardware from the same supplier) requires the ICT department to think ahead and choose the right products, including applications packages or system-software products.

- *Staff commitment will improve* if a clearly defined ICT strategy is in place. If systems staff believe that no attempt is being made to improve the development environment or use up-to-date techniques, they may feel that their career objectives would be better served in a more forward looking organization.

In a rapidly changing business and technical environment, the ability to respond rapidly to new business requirements can be greatly facilitated by the development of an ICT strategy, incorporating appropriate ICT standards. This view was supported by the results of a survey published by CSC Index (1989). Without an ICT strategy that includes a flexible hardware and software architecture, there will inevitably be delays in implementing applications to meet new business requirements.

THE VALUE OF ICT STANDARDS DEPENDS ON HOW THEY ARE USED TO SUPPORT THE ORGANIZATION'S INFORMATION NEEDS

The aim of an ICT strategy, as for any other business strategy, is to create a long-term plan for achieving success. This implies that the strategy must be aimed at achieving well-defined goals. The strategy is then implemented by managing all of the organization's ICT resources (mainframes, personal computers and workstations, communications networks, systems software, applications developed in-house, application packages, bought-in software services, in-house software support database management systems, and so on) to achieve those aims in a changing and competitive business environment.

The terms in which ICT-strategy goals are expressed are determined largely by the way in which an ICT department perceives its basic mission. An inward-looking department is more likely to express ICT-strategy goals in technical terms, and will aim to achieve those goals by managing its own resources (hardware, systems staff, existing software, and so on).

The problems that can result from basing an ICT strategy purely on technical goals are illustrated by the experience of an insurance company, which is summarized in Figure 1. By failing to develop a close relationship with its users, the systems department of this organization was unaware of the business goals, and thus directed all its attention to achieving its own technical goal of providing a sound software base. Although this may well have been needed, the user community could not relate the large expenditure on software to any direct benefits in terms of its own business strategy. The result was that the systems department's budget was reduced substantially, and its only possible goal was then to concentrate on surviving.

This clearly illustrates an important point, that expenditure on ICT should be in line with the *business* benefits that will be achieved. For this reason, many public sector organizations have had to rethink their ICT strategies as a consequence of the changes in operating conditions brought about by government legislation.

The situation is very different if the ICT strategy goals are expressed in business terms. In this case, the ICT department is outward looking, perceiving its main mission to be that of helping the organization achieve its business goals. In turn, the organization perceives IT as just, one of the business resources that needs to be managed in order to achieve those goals. The barriers to achieving the goals are determined by the business and commercial environment in which the organization operates - its competitors, the political and economic environment, legal and regulatory constraints, and so on. Thus, IT is perceived as making a real contribution to achieving business goals, and the ICT strategy is an integral part of the business strategy.

Consequent upon this, can come a change in the perception of the cost of technology in relation to the value of information. When organizations seek to define the return on an investment in ICT, they tend to concentrate on the cost of the technology and to ignore the value of the information. Few ask what return they get from information, yet it is the information that is critical. The technology that supports the information is of value only insofar as it allows better use to be made of information. The search for proof that investment in ICT provides business value will continue to be futile unless management - both business and systems - recognizes that divorcing the cost of the technology from the value of the information it supports is counterproductive.

Thus, the value that an organization gets from ICT depends on how well it manages its investment in *technology* so as to maximize the return it gets from information. The ability of the organization to manage the technology in relation to the organization and its information needs is critical to successful investment in ICT. Investing in ICT to maximize the benefits is an organizational issue first, and a technology issue second.

In general, ICT-strategy goals should therefore be expressed in terms of

---

An insurance company

This insurance company (which wishes to remain anonymous) employs about 7,000 people, 350 of whom work in the information systems department. All of the computing resources are supplied by IBM, and the ICT strategy is technology-driven, being determined largely by IBM's current products and future product developments. Most development work is done in-house, using IBM products for the ICT infrastructure. The systems department is not represented at main-board level and communication between the department and top management is poor. There is no mechanism for systems staff to feed ideas upwards and, although there is some downward communication of business strategy, many of the most important business decisions are not relayed to the systems department.

As a consequence, the systems department is not well regarded by the user community, which feels that its needs are not understood and not being properly met. Not surprisingly, user departments are beginning to acquire their own computing resources, and are increasingly reluctant to pay for the systems department's services. As a result, the department's budget has been severely cut, and its staff has been reduced by about 25 per cent. These reductions mean that the systems department is now able to operate only in 'fire fighting' mode, and its ability to support users properly is declining still further. The department is caught in a vicious circle from which it can see no means of escape.

Figure 1 Defining an ICT Strategy in Terms of Technical Goals Can Lead to Serious Problems

helping the 'business' functions achieve the organization's goals. For example, online-ordering and stock-availability systems can help the sales department of a distribution company to provide a better service to customers and thus increase sales. Similarly, in a large multinational group such as a car manufacturer or airline company, or national public service like the National Health Service, a corporate-wide data standard can help the organization share vital information and provide an efficient service.

STANDARDS SUPPORT DIFFERENT TYPES OF ICT INVESTMENT

There are many ways of classifying information systems. A technology-oriented classification would be data processing systems, office systems, and telecommunications systems. An application-oriented classification would be sales and marketing systems, financial and accounting systems, management information systems, and so on. The most appropriate classification for investment-appraisal purposes, however, is the *business purpose* of the proposed investment.

As Figure 2 shows, there can be five main business purposes for investing in

ICT (Willcocks, 1992), which can lead to five types of ICT investment:

- *Mandatory investments*. These are the investments that the organization must make because of commercial or statutory pressures. Most organizations recognize that they have no choice but to invest in some kinds of systems in order to survive and to operate legally and effectively. Research indicates (CSC Index, 1990) that in some organizations, as much as 80 or 90 per cent of ICT expenditure may be mandatory. Investment in mandatory systems is sometimes called 'the threshold investment' in ICT; and represents the amount of money that an organization must invest in ICT if it is simply to survive. Because the organization has no choice but to invest in mandatory systems, the main investment consideration is how the total costs of the system can be minimized and which available design option will be most cost-effective.

- *Investments to improve business performance* . These are investments that are aimed at improving the organization's business performance by reducing costs, or increasing revenues. Most commercial organizations aim to achieve growth in revenue and profitability. Greater profitability can be achieved by either reducing costs or increasing revenues, and information systems may contribute to either in various ways, as shown in Figure 3.

The role of systems in reducing costs is well established and cost reduction continues to be an important criterion for justifying ICT investment proposals. However, it is not always easy to apply conventional cost/benefit analysis to ICT investment proposals. For some types of benefit (cost reductions, projected increases in sales, or reductions in staff, for example), the monetary value of the benefits can be estimated with a high degree of certainty. Other types of benefits,

Purpose of Investment	Type of ICT Investment
Survival and functioning	Mandatory investments
Improving performance by reducing costs or increasing revenue	Investments to improve performance
Achieving a competitive leap	Competitive edge investment
Enabling the benefits of other ICT investments to be realised	Infrastructure investments
Being prepared to compete effectively in the future	Research investments

Figure 2 The organisational purpose defines the category of ICT Investment

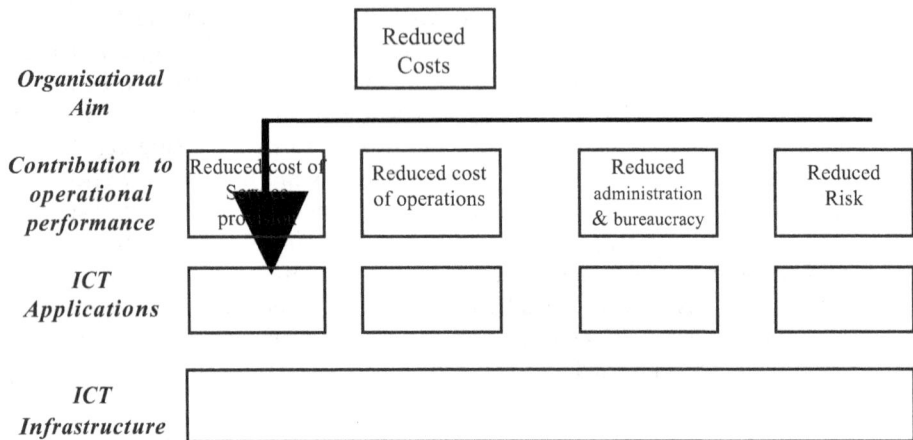

```
                        ┌──────────────┐
                        │   Reduced    │
Organisational          │    Costs     │
     Aim                └──────────────┘
```

Figure 3 Relating ICT to organisational objectives

such as better decisionmaking or improved interpersonal communications, are
difficult to quantify. In addition, the link between the investment and the benefit
may be indirect, and therefore hard to verify.

In practice, most ICT investment decisions involve both an assessment of the
cost and benefits (expressed in monetary terms), and management judgement. The
more subjective the assessment of benefits is, the greater the need for management
judgement. For instance, the 'cost-displacement' method is relatively straightforward to use and verify because it uses objective measurement criteria. Other
methods, such as the 'break-even method', can be highly subjective because they
rely on managers using their judgement to place a value on the system.

- *Investments to gain competitive advantage.* These are aimed at improving the
  organization's share of, or position in, its market. Evaluating such proposals,
  however, is more than simply assessing the extent to which revenues or
  profitability may be increased. A company that is a market leader in its
  industry may decide, for example, to forego short-term profits to achieve such
  a leap. Such a company can exercise significant control over the pricing and
  cost structure of the industry and its future direction. ICT investment proposals should be evaluated in the light of the long-term business advantages that
  may accrue.
- *Architecture investments.* These are investments in the technical facilities
  needed to support business applications. They do not offer direct benefits, but
  enable the benefits of other ICT investments to realized. For many organizations, justifying the fourth type of ICT investment is a major problem, as
  investments in ICT infrastructure enable the benefits of other types of ICT
  investment to be realized.

It will always be particularly difficult to justify investment in the ICT infrastructure just as it is to justify infrastructure investments in other areas of the
business (office buildings, telephones, and so on). Two main problems tend to arise
in justifying ICT infrastructure investments:

- The benefits of infrastructure systems accrue to the organization as a whole and are not always readily visible in the short-term performance of individual units or functions. These units are, however, judged by their operational performance and their managers are rewarded and promoted on the basis of that performance. This makes it difficult for them to accept an organization-wide view of benefits, and they can be reluctant to sponsor the cost of such investments.
- As it may take a long time to develop and implement architecture/infrastructure systems, no return may be achieved on the investment until well into the future. Infrastructure requires standardization of ICT, including data structures, security, systems maintenance methods and so on. Most managers of regional and local units are concerned with achieving short-term (business) goals.

Ultimately, central management (at the relevant level) must form a judgement as to whether the potential benefits of the proposed infrastructure justify the investment. However, the justification process for ICT architecture investments can be made easier by:

- Establishing performance standards for the ICT infrastructure.
- Including a share of architecture costs when evaluating application systems.

Corporate culture also plays an important role in determining how infrastructure investment proposals are considered, as is well illustrated by Sweden Post. Investing in infrastructure and applications is clearly focused on developing business opportunities. The purpose of its ICT strategy is to support the business operations and to create the conditions for a high degree of freedom in choosing application areas for ICT.

- *Investments in research projects*. These are made with the aim of ensuring that the organization is prepared for the future so that it can continue to sustain or improve its competitive position. Allocating a predefined amount of money for research projects, and setting clear objectives and budgets for the projects, is the common way of funding such work. However, the business benefits of research projects usually take several years to become evident. Hence, how much to spend on researching future applications and products is a question of judging the future needs of the business, and the value of individual research projects in preparing the organization to meet those needs.

EVALUATING THE CONTRIBUTION OF ICT STANDARDS

There is no single 'methodology' for proving the contribution of ICT standards to business performance. Nor can there be (as many authorities agree e.g. Parker et al (1989), Remenyi et al (1991), Farbey et al (1993)). Because the value of ICT is inextricably linked with an organization's ability to exploit its information asset by using technology, putting a precise value on the contribution of standards alone would be misleading. As in other parts of the business, it can be difficult to isolate the precise business contribution of individual investments.

Developing Appropriate Business Measures

The preceding remarks do not mean, however, that no measurements should be made. Senior management and users in general look for evidence that ICT is as well managed as any other aspect of the business. Monitoring performance in line with appropriate measures is one aspect of good management. The set of business performance measures to which ICT investment in standards can be related and which we describe below is designed to meet this purpose.

A set of ratios can be used to monitor the relationship between ICT expenditure and overall business performance (CSC Index, 1990) . The ratios relate ICT expenditure to the four main measures of business performance with which managers are already familiar:

- *Size:* For commercial organizations, size can be measured in terms of revenue (or its equivalent, such as patient income for hospitals or general practitioners) or number of employees. Thus, two of the ratios will be ICT expenditure as a proportion of revenue, and ICT expenditure per employee.
- *Business volume:* This is the volume of business carried out, measured in terms other than money. It applies to both commercial and public-sector organizations. Examples include the population served by a local hospital trust, the total number of passengers carried by an airline, or the total generating capacity of an electricity company. Thus, in a hospital trust, one of the ratios would be ICT expenditure per head of population served, and in an airline, ICT expenditure per passenger carried.
- *Operating expenses* (including ICT costs). Another ratio would be ICT expenditure as a proportion of operating expenses.
- *Key business indicators:* These measure business performance in nonfinancial terms. The common basis for such indicators is to measure business activity or volume in terms of resources employed (bed occupancy factors for a hospital).

Using the framework of the four types of business-performance measures listed above, each organization needs to establish its own unique set of ratios that relate ICT expenditure to its key organizational measures and indicators. It can be helpful to relate ICT expenditure to more than one measure for each category of performance to reflect the different activities and objectives of the organization. However, it is important to avoid too many.

Each of the ratios can be tracked over time to build up a composite picture of the contribution of ICT to overall business performance, and to smooth out fluctuations in business performance and any unusually high or low ICT costs, or high or low use, in a particular period. Three years is probably the minimum length of time needed by most organizations to get a reliable picture of the main trends. The level of utilization of standards is one factor which could be monitored.

The primary purpose of the ratios is to establish whether the general picture is a healthy one. They are therefore intended to be an aid to management judgment —a diagnostic tool, not an absolute 'proof' of the value of standards in ICT. If the trends indicate cause for concern, a more detailed analysis can be carried out when it is appropriate.

The advantage of using a set of ratios that relate ICT expenditure to business-performance measures is that it allows the contribution of ICT to be assessed in a holistic manner. The use of such indicators can be particularly helpful in the public sector, as they provide a measure of the contribution of ICT in relation to operational volumes and not just in terms of operating costs. In a consultancy project conducted by the author, it was suggested that ICT expenditure per patient discharged from a hospital may be an appropriate indicator for the UK National Health Service (NHS).

If the organization is highly decentralized or engages in a wide range of diverse operations (e.g. conglomerates), the framework of business-performance measures that we have described can be used to establish a set of ratios for each local unit, as well as for the organization as a whole.

Manage 'Outcomes' To Achieve Benefit Delivery

The approach described above has been advocated in a more general context by Ward (1994). He suggests that organizations should adopt a benefits management approach to complement ICT project processes. Such an approach requires a comprehensive documented process for identifying and managing benefits, including the allocation of responsibility for identification and delivery of the benefits of ICT investment.

An interesting aspect of this approach is that it explicitly recognizes that ICT investments inevitably involve change, and can produce effects which may be positive or negative. Equally, the effects of some changes can be predicted, while others cannot. Such unexpected changes often arise as a result of human reaction to a new system. Ward uses the term 'outcomes' of ICT investments (which will always occur) rather than benefits (which may not occur).

Figure 4 illustrates the four aspects of the resultant benefits management matrix, consideration of which can help an organization develop its ability to improve benefits delivery. The matrix is valid for a single project or a range of projects. Positive expected outcomes, expressed in the investment justification, should be achieved by good management practice. Positive unexpected outcomes are a bonus. Negative expected outcomes are "the price worth paying" for the changes and can be managed as part of the change process. Negative unexpected outcomes are not predicted in advance! However, risk assessment processes can be used to help forecast possible problems.

Both the use of measures and the definition of outcomes have been incorporated into a framework for standards development described by Moreton et al (1995). This framework recognizes that standards development is a very complex issue covering a wide range of user requirements and associated technical areas. Obviously, few departments or project teams work in total isolation, so there is a need to specify standards for specific projects and also to ensure interworking between the systems developed. It is unrealistic to hope for a single standard that covers all technical areas. Because technology is advancing at a rapid rate, organizations will always be in a situation where new standards are emerging that may need to be adopted by that organization. Thus any framework to aid the selection of standards must take account of the temporality of technology and associated standards, and the variety of standards in a wide range of contexts.

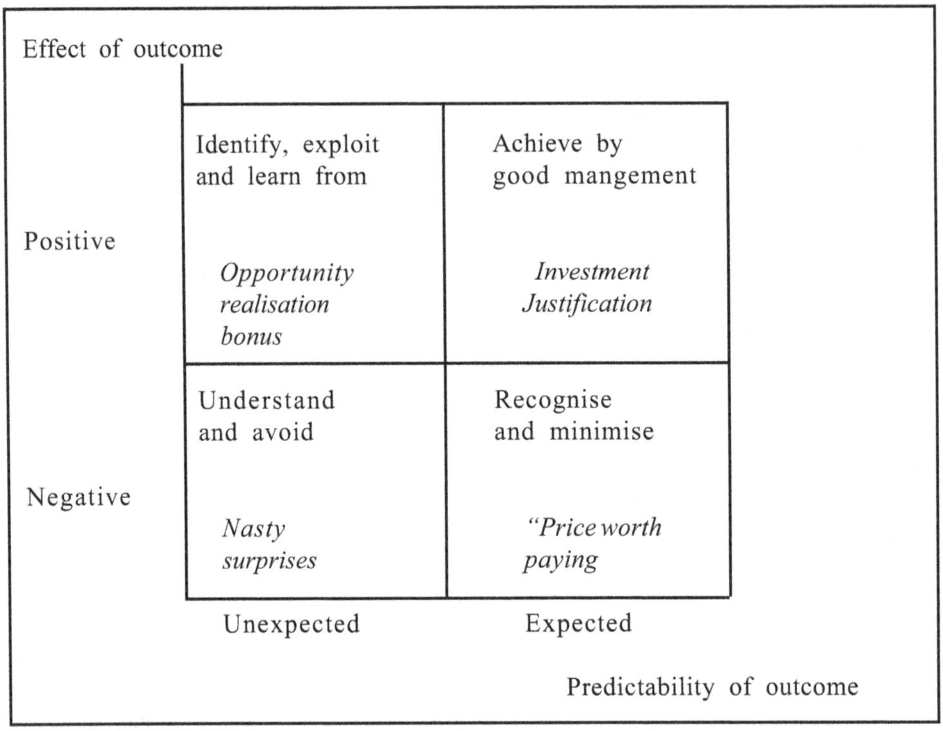

Effect of outcome		
Positive	Identify, exploit and learn from *Opportunity realisation bonus*	Achieve by good mangement *Investment Justification*
Negative	Understand and avoid *Nasty surprises*	Recognise and minimise *"Price worth paying*
	Unexpected	Expected

Predictability of outcome

Figure 4 Benefits Management Matrix

CONCLUSION

In this chapter we have stressed that ICT strategy, architecture and standards must relate to business benefits. We have argued that the value of ICT standards depends on how they are used to support the organization's information needs. A set of ratios were proposed which can be used to monitor the relationship between ICT expenditure and overall organizational performance. Additionally, a broader benefits management approach was advocated, in which ICT investment outcomes can be managed to achieve benefit delivery for ICT standards. Both the use of measures and the definition of outcomes can be incorporated into a framework for standards development.

From this discussion, it is clear that there is a need for a coherent approach to the selection, use and evaluation of ICT standards to support an organization's business and ICT strategies. The approach must address the following issues:

- The need to establish requirements for ICT standards in terms of both the business goals that need to be achieved and the technical requirements that underpin these. In any project, the application of standards must be done to realize business as well as technical benefits.
- Specification of standards to meet both long-term strategic and project-specific (or local) needs in an organizational context in which ICT strategy,

architecture, infrastructure implementation and procurement can be carried out at several levels. In this environment, standards development at a particular level is likely to be influenced by ICT strategies developed by a number of other organizational units both within and outside of the organization. For example in the Health Service, a hospital trust developing a (local) ICT strategy, architecture and infrastructure is influenced by the national architecture which encompasses NHS standards.

- The need to procure or build products which conform to the architecture defined by the organizational unit or project. In some cases standards are simply used directly without any procurement or product building (e.g. documentation and data standards for application development).

- The need to evaluate the effectiveness of standards in contributing to the achievement of stated business goals.

REFERENCES

CSC Index (1989). Software strategy. Foundation Research Report 69. CSC Index.

CSC Index (1990). *Getting value from IT*. Foundation Research Report 75. CSC Index.

Earl M (1989), *Management Strategies for Information Technology*, Prentice Hall.

Farbey B, Land F and Targett D (1993). *How to Assess Your IT Investment*. Oxford Butterworth-Heinemann. ISBN 0750606541.

Moreton R. (1989). Managing data in a complex systems environment. *Journal of Information Technology*, 4, 1, pp 49-54.

Moreton R, Sloane A and Simon E (1995). Implementing Information Management and Technology Standards: a framework. *Technology Management*, 2, 6, pp 275-288.

Parker M, Trainor H E and Benson R J. (1989). *Information Strategy and Economics: Linking Business Performance to IT*. Prentice Hall.

Peppard J (1993). I.T. *Strategy for Business*, Pitman.

Remenyi D S J, Money A and Twite A (1991). *A Guide to Measuring and Managing IT Benefit*. Oxford Blackwell.

Robson W (1997). *Strategic Management & Information Systems*. London Pitman.

Ward J, Griffiths P, Whitmore P (1997). *Strategic Planning for Information Systems*, Wiley.

Ward J, (1994). Information Systems - Delivering Business Value?. *Proceedings 4th Annual BIT Conference*, November, 8-19.

Willcocks L, (1992). Evaluating IT investments: research findings and reappraisal. *Journal of Information Systems*, 2, 243-268.

Chapter VIII

The Role of Standards for Interoperating Information Systems

Wilhelm Hasselbring
INFOLAB, Tilburg University

INTRODUCTION

For integrating heterogeneous information systems, semantic interoperability is necessary to ensure that exchange of information makes sense — that the provider and requester of information have a common understanding of the 'meaning' of the requested services and data. Effective exchange of information between heterogeneous systems needs to be based on a common understanding of the transferred data. This chapter discusses the role of domain-specific standards for managing semantic heterogeneity among dissimilar information sources. The *process* of integrating such heterogeneous information systems is also discussed in this context, whereby standards play a central role for 'initiating' *top-down* processes by means of defining common data models for the involved information sources.

BACKGROUND

Traditionally, the integration of heterogeneous information systems proceeds in a *bottom-up* process. Information stored in existing legacy systems is analyzed with respect to potential overlaps, whereby overlapping data in dissimilar systems describes the same or related information. The overlapping areas of related information sources are subsequently integrated. The integration is usually realized by means of mediators, federated database systems or such—like system architectures. Typical goals for the integration of existing systems are the development of global applications that access the data from multiple sources as well as consistency management of information that is stored in related systems.

Let us consider digital libraries as an example domain where the integration of existing information sources is one of the central problems to be solved (Schatz & Chen, 1996). As one result of a bottom-up integration of those existing information sources, the structure of the merged common data model (schema) is determined by the overlaps among the local data models, and not by the requirements of global applications. The maintenance of such integrated models is a problem, because those merged models rapidly become very complex; usually more complex than required for the actual integration goals. This situation can lead to severe scalability problems with respect to execution performance, usability and maintenance.

ISSUES, CONTROVERSIES AND PROBLEMS WITH THE TRADITIONAL BOTTOM-UP INTEGRATION PROCESS

Various approaches for integrating heterogeneous information systems — e.g., federated database systems or mediator and agent architectures — have been proposed (Hurson, Bright & Pakzad, 1993, Elmagarmid, Rusinkiewicz & Sheth, 1998, Sheth, 1998, Wiederhold, 1996, Jennings & Wooldridge, 1998). We illustrate the traditional bottom-up process of integrating such heterogeneous information systems, by means of schema integration in federated database systems (Sheth & Larson, 1990). A federated database system is an integration of autonomous database systems, where both local applications and global applications accessing multiple database systems are supported. For federated database systems, the traditional three-level schema architecture must be extended to support the dimensions of distribution, heterogeneity, and autonomy. The five-level-schema-architecture of Sheth & Larson (1990) is generally accepted as the basic structure for schema integration in federated database systems or at least for comparison with other architectures of specific federated database systems (Conrad, Eaglestone, Hasselbring, Roantree, Saltor, Schönhoff, Strässler & Vermeer 1997).

Figure 1 illustrates the bottom-up process for constructing the schema architecture in federated database systems which starts with an analysis of information stored in the local systems. To explain the schema types displayed in Figure 1: A *local* schema is the conceptual schema of a component database system, which is expressed in the (native) data model of that component database system. In a first step, the local schemas are translated into component schemas in the canonical data model of the federation layer. Then, export schemas are filtered from the component schemas and merged into a common federated schema. A *component* schema is a local schema transformed into the so-called *canonical* data model of the federation layer. The component, export and federated schemas are defined in this *canonical* data model. An *export* schema is derived from a component schema and defines an interface to the local data that is made available to the federation. A *federated* schema is the result of the integration of multiple export schemas, and thus provides a uniform interface for global applications. An *external* schema is a specific view on a federated schema or on a local schema, which serves as a specific interface for applications (local or global). To keep it simple, no external schemas are shown in Figure 1. For local applications, external schemas can be filtered from the local schemas. For global applications, external schemas are filtered from the federated schema.

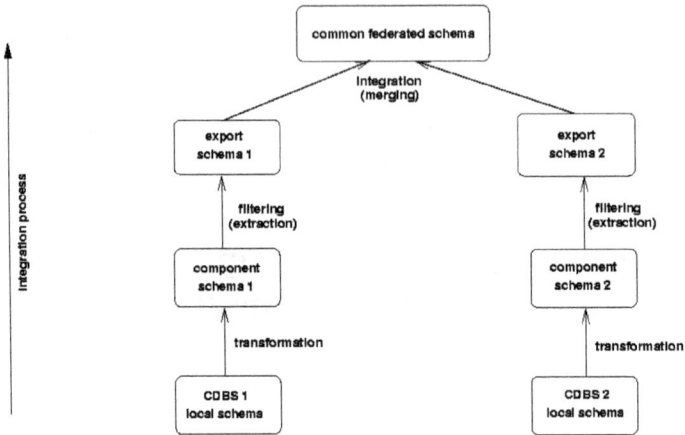

Figure 1: Bottom-up integration on the schema level. The arrows illustrate the development process that starts with the local data models of some existing legacy systems.

The translation from local to component schemas eliminates the data model heterogeneity among the various local schemas. Since the component and export schemas are provided in a canonical data model, no model translation is needed on the federation layer. However, many discrepancies between the export schemas may exist. For example, similar entities may be specified at different abstraction levels, or equivalent attributes may have different data types.

As one characteristic result obtained by a bottom-up construction of the schema architecture, the structure of the common federated schema is determined by the overlaps among the local schemas. Those schemas *overlap* when the intersections of the corresponding extensions are not empty. For instance, most object-oriented integration approaches resolve semantic overlappings by introducing generalized classes such that the original inheritance hierarchies are sub-hierarchies of the resulting merged hierarchy (upward-inheritance principle) (Dayal & Hwang, 1984, Schrefl & Neuhold, 1988). These approaches take over the inheritance hierarchies from the local schemas to the integrated schema level and adapt them to each other. Consequently, the resulting merged hierarchies become needlessly complex.

We illustrate these problems by means of a small example for bottom-up integration of two digital library databases (this example is adopted from Schmitt & Saake (1998)). The first local database to be integrated is a library database storing information about publications, where books and journal papers are special types of publications (see Figure 2(a)). The second local database is a project publication database which stores information about publications related to a project. In the project database, books and technical reports are considered non-refereed publications (see Figure 2(b)).

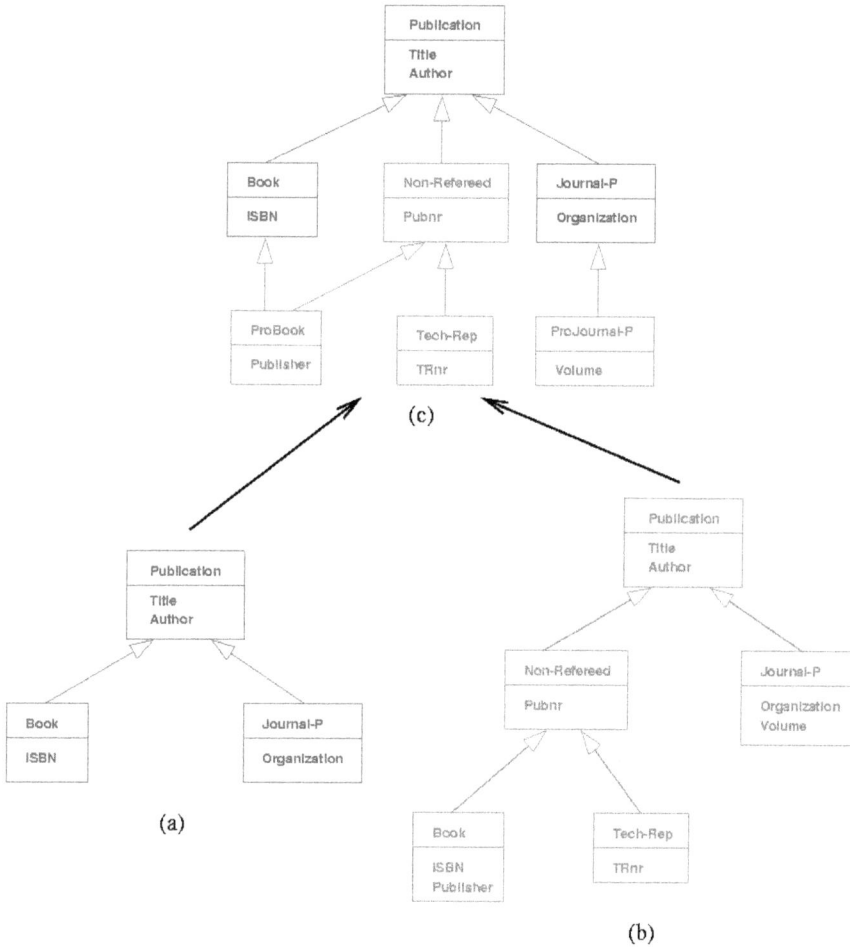

Figure 2: Local schemas of the library (a) and project publication (b) databases as well as the integrated schema as a result of merging (c). The schemas are modeled in the UML notation for class diagrams (Booch, Rumbaugh, & Jacobson 1999).

Integrating these local databases requires an analysis of the extensional overlappings among the classes to be integrated, whereby the *extension* of a class represents the set of *possible* objects. For this example, we assume that the Publication extensions of both databases overlap such that the corresponding merged class' extension represents the union of the local extensions. Some publications in the project publication database are published simultaneously as a technical report and as a journal paper. For simplicity, we assume that attribute conflicts are already resolved, i.e., attributes with identical names have the same semantics.

By means of a decomposition of the extensional overlappings into base extensions and their subsequent integration, an integrated schema (Figure 2(c)) with seven classes can be constructed, whereas the project publication database contains only five classes and the three classes for the library database are almost identical to a subset of the project's classes. Due to the upward-inheritance principle, the integrated schema contains (approximately) the sum of the classes contained in the individual component schemas. For reducing this complexity within the integrated schema, classes may be merged (Schmitt & Saake, 1998):

Vertical merging means merging a class with its direct superclass. In Figure 2(c), the classes Journal-P and ProJournal-P could be merged vertically. This vertical merging would change Volume into an *optional* attribute within the merged class (in the local databases all attributes are mandatory, i.e., NULL values are not allowed).

Horizontal merging means merging classes that have a direct specialization relation to the same superclass. In Figure 2(c), the classes Book and Non-Refereed could be merged horizontally. This horizontal merging would change both Pubnr and ISBN into *optional* attributes within the merged class; thus again introducing optional attributes. Furthermore, the additional integrity constraint would be introduced, which requires at least one attribute in the merged class holding a valid value.

To summarize: This small example illustrates the fact that with the traditional bottom-up process, merged class hierarchies are usually more complex than the integrated local schemas. Optimizing these merged hierarchies with respect to minimizing the number of classes may introduce new constraints; thus, complicating the common federated schema. These problems are largely due to the upward-inheritance principle, i.e., the original inheritance hierarchies are sub-hierarchies of the resulting merged hierarchy. New integrity constraints and optional attributes are often introduced.

Under those traditional integration paradigms (Batini, Lenzerini & Navathe, 1986), the integrated view/schema depends directly on the source schemas. An integration engineer defines the desired integrated view by examining all the existing systems to be integrated. The bottom-up approach is to sum up the capacity of existing information systems in one global model. As a result, the usability and maintainability of such integrated schemas can become a serious problem.

SOLUTIONS AND RECOMMENDATIONS

To approach a more 'ideal' top-down integration process, we start with a look at domain-specific software development before the top-down integration process is discussed. Finally, a combined 'yo-yo' approach is discussed in this section. The appendix presents a list of some domain-specific standards that are relevant for interoperability of information systems.

Domain engineering

Domain engineering is an activity for building reusable components, whereby the systematic creation of domain models and architectures is addressed. Domain engineering aims at supporting *application engineering* which uses the domain models and architectures to build concrete systems. The emphasis is on reuse and product lines. The Domain-Specific Software Architecture (DSSA) (Taylor, Tracz & Coglianese, 1995) engineering process was introduced to promote a clear distinction between domain and application requirements. A Domain-Specific Software Architecture consists of a domain model and a reference architecture as modeled in the blue part of Figure 3. The DSSA process consists of domain analysis, architecture modeling, design, and implementation stages as illustrated in Figure 4.

Domain models represent the set of requirements that are common to systems within a specific domain. Usually, those systems can be grouped into product lines (Dikel, Kane, Ornburn, Loftus, and Wilson, 1997), for instance, for the insurance or banking domain. There may be many domains, or areas of expertise, represented in a single product line and a single domain may span multiple product lines. *Domain analysis* is the process of identifying, collecting, organizing, and representing the relevant information in a domain, based upon the study of existing systems and their development histories and knowledge captured from domain experts. Figure 3 illustrates the relations between some roles (domain experts and application engineers) and the artifacts in the DSSA engineering process.

The *architecture* of a software system defines that system in terms of components and interactions/connections among those components. It is not the *design* of that system which is more detailed. The architecture shows the correspondence between the requirements and the constructed system, thereby providing some rationale for the design decisions (Shaw & Garlan, 1996). Reference architectures are the structures used to build systems in a product line. The domain model characterizes the *problem space*, while the *reference architecture* addresses the *solution space* (design). The *reference requirements* within the domain model define the (generic) functional requirements for applications in a domain.

In *application engineering*, a developer uses the domain models within the product line to understand the capabilities offered by the reference architecture and specifies a system for development. The developer then uses the reference architecture to build the system. An architectural model is developed in this phase from which detailed design and implementation can be done. Application engineers use the domain models with the users to elicit information about particular systems and to define the requirements for the planned software systems. By so doing, the models frame the user's needs in terms of existing models. Those needs not covered by a domain model are new requirements. Once a tentative set of features have been identified, the engineers analyze the interaction among the features to assess feasibility and identify additional requirements and contexts not described in a domain model. The domain engineers may choose to update a domain model with the new requirements. As indicated by the dashed arrows in Figure 4, various forms of feedback are possible.

Domain engineering and application engineering are complementary, interacting, parallel processes that comprise a reuse-oriented software production. An application engineering process should develop software systems from reusable components created by a domain engineering process (see Figure 4). The focus of application engineering is a single system whereas the focus of domain engineering is on multiple related systems within a domain. Typical application engineering

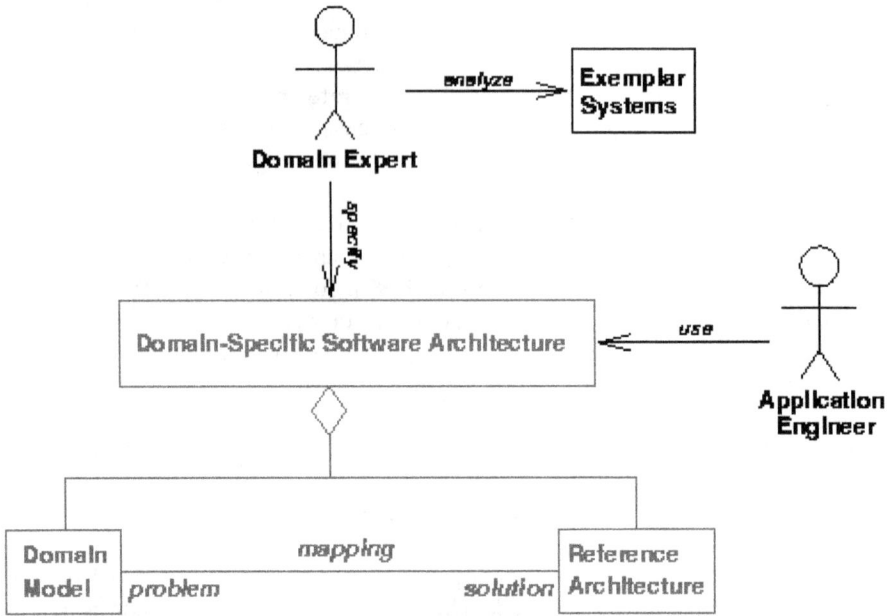

Figure 3: Relations between some roles and artifacts in the DSSA engineering process. Hollow diamonds indicate part-of relations. We use the UML notation for actors to model the roles (Booch et al., 1999).

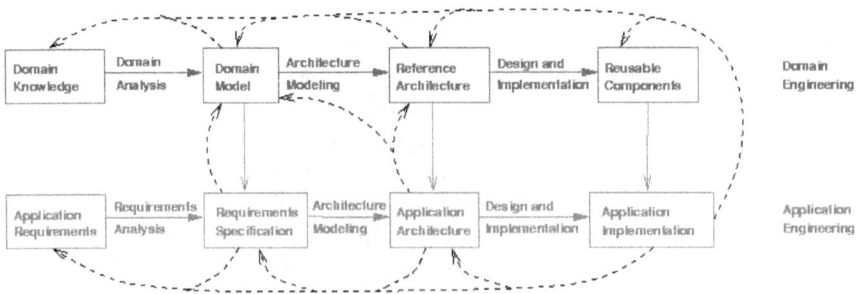

Figure 4: The DSSA engineering process. In application engineering, software systems are developed from reusable components created by a domain engineering process. As indicated by the dashed arrows, various forms of feedback are possible.

activities include using a domain model to identify customer requirements, and a (generic) reference architecture to specify an application architecture. Domain engineering emphasizes on the principle of *design-for-reuse* whereas application engineering should be based on the principle of *design-with-reuse*.

Top-Down Integration

Despite the fact that engineering of (new) global applications will usually require the integration of existing information sources, we argue that the integration process should proceed in a top-down way starting with the data models that are common to all the involved local systems, i.e. with domain models in the context of a DSSA engineering process. For such a *top-down* integration of those heterogeneous information systems, we propose the use of domain-specific standards as the basis for the common data models. Some relevant standards are listed in the appendix.

Figure 5 illustrates the top-down process for constructing the schema architecture which starts with the common federated data schema that should be based on requirements of global applications and on relevant standards. The individual local schemas are integrated into this common schema via the component and export schemas. Export schemas are filtered from the common federated schema, instead of merging them into the common federated schema, as illustrated in Figure 5. Afterwards, these export schemas are mapped to the component schemas, which are transformed into the native data models of the local systems.

To 'hook' a local information system's model into a common (domain-specific) model, the responsive integration engineer has to understand his or her local information system and the corresponding domain model, but does *not* have

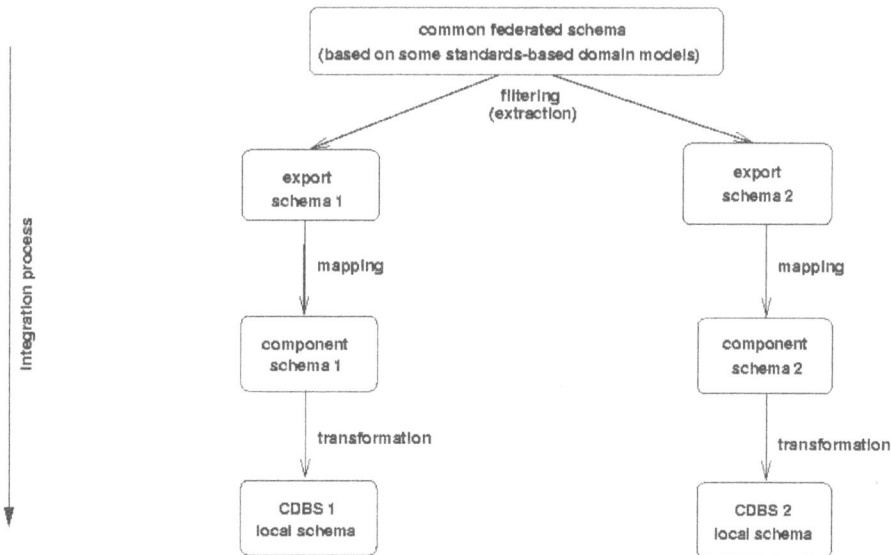

Figure 5: Top-down integration on the schema level. The arrows illustrate the development process that starts with the common standards-based data schema.

to understand the other local information systems and the constraints that could be introduced by them into the common model using a bottom-up process. In the domain of digital libraries, for instance, Z39.50 and the Dublin Core can be the basis for the common models (see the appendix). The approach is to define a common model based on the information we want to be in it. Then we map to the relevant bits of data in the local information systems. This should result in workable approach.

A Combined Yo-Yo Approach

In practice, we can also expect a *yo-yo* approach as illustrated in Figure 6: the integration process alternates with bottom-up and top-down steps. A top-down process can be expected to support global applications — such as workflow management or decision support functions to be supported by a data warehouse. A bottom-up process can be expected in the case that some local systems regularly need information from other systems — e.g., if one library system needs citation information from another library system.

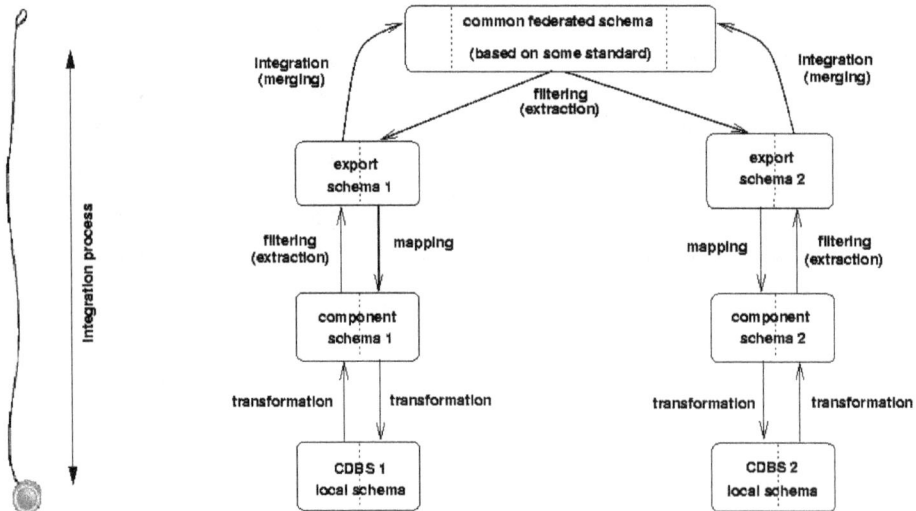

Figure 6: A combined yo-yo *approach to exploit the benefits of both strategies.*

The bottom-up process may provide input for extending the domain-specific standards as indicated in Figure 4. Then, a method for new information which resides in a local information system, but was not originally intended in the common model, is required. It should be easy to update the common model with such a 'yo-yo' approach. The evolution of common ontologies for mediation of information sharing is discussed in Kahng & McLeod (1998).

CONCLUSIONS

Some problems with the traditional bottom-up approach to the integration of heterogeneous information system have been discussed and a 'ideal' top-down approach using domain-specific standards to achieve more usable and scalable software architectures for heterogeneous information systems is proposed. The bottom-up process starts with an analysis of information stored in existing legacy systems to explore potential overlaps among information stored within the individual components. The top-down process starts with a common data model that should be based on domain-specific standards.

Enabling existing legacy systems to exchange information will typically start at the bottom and designing new global applications with access to several local systems will typically start at the top. Anyway, engineering of (new) global applications will usually require the integration of existing information sources. Despite this fact, we argue that the integration process should proceed in a top-down way starting with the data models that are common to all the involved local systems, i.e. with standards-based domain models. The combined *yo-yo* approach aims at exploiting the benefits of both strategies and to serve as a migration path form the traditional bottom-up approach towards an ideal top-down approach.

Usually, one result of the traditional bottom-up approach to the integration of heterogeneous information systems is that the structure of the common, merged models is determined by the overlaps among the local models. As opposed to the bottom-up approach, with the top-down approach the structure of the common model is *not* determined by the overlaps among the local models, but by the requirements of global applications and domain-specific standards, which can become the basis for *semantic* interoperability.

Particularly, the use of domain-specific standards as the basis for the common data model should alleviate the integration of commercial components that offer standards-compliant interfaces. For such a *top-down* integration of those heterogeneous information systems, we propose the use of domain-specific standards as the basis for the common data models. The use of such standards within the integration system also encourages a (smooth) migration towards modern standards-compliant systems. With standards-compliant interfaces, it becomes straightforward to 'hook' the local information of these subsystems into the common model. To be successful, it is obvious that both the standards and the local applications must be in the same domain. In any case, the selected domain-specific standards must cover the application domains of the integrated information systems. Only extracts of the standards will be used for actual applications. Common models should be restricted to specific domains (Missier, Rusinkiewicz & Silberschatz, 1995).

To quote Kleewein (1996) on practical issues with commercial use of federated databases:

"Schema integration is one aspect of usability that impedes federation. There are often thousands of tables or views involved in a federation making maintenance of a global schema difficult."

Starting with the global schema and basing it on standards avoids changes to the fundamental structure of this schema making integration more usable and scalable. Only when interoperability is based on standards will this be realizable. Appropriate standards may enable interoperability, for instance Z39.50 and the Dublin Core in the domain of digital libraries. To quote Paepcke, Chang, Garcia-Molina & Winograd (1998) on the relevance of standards for interoperability for

digital libraries worldwide:
> "An appropriate standard that is widely adhered to provides a powerful interoperability tool."

However, it is not easy to find those domain-specific standards that are detailed enough to allow for real interoperability. This is an important area for future work.

Appendix A: Some Domain-Specific Standards

This list of some relevant domain-specific standards is not meant to be comprehensive, but representative. The standards are grouped into domains and within groups the standards are listed in alphabetical order. References to corresponding web pages are provided. Exemplary, standards for the domains of digital libraries, healthcare, manufacturing, enterprise application integration, and electronic commerce are presented.

Digital Libraries

Dublin Core The Dublin Core is a metadata element set intended to describe electronic information sources. Originally conceived for author-generated description of Web resources, it has attracted the attention of formal resource description communities such as digital libraries.
More information is available at http://purl.org/dc/

Z39.50 The Z39.50 standard for searching library catalogs is a standard for the retrieval of bibliographic records independent of the type of system on which they are stored. The Z39.50 standard is an open systems protocol established and maintained by the library community. This standard is the basis for open systems connectivity among different library systems and information sources.
More information is available at http://lcweb.loc.gov/z3950/agency/

Healthcare

CEN TC 251 The scope of Working Group I (Information Models) of the Technical Committee 251 for Health Informatics within the European Committee for Standardization is the development of European standards to facilitate communication between independent information systems within and between organizations, for health related purposes, e.g., blood transfusion, physiology, pharmacology, psychiatry and nursing. The standards are based on information models - generic models of aspects of health care or health care information. The domain of this work are standards for electronic medical records and messages to meet specific healthcare business needs for the communication of healthcare information.
More information is available at http://www.centc251.org/

HL7 The Health Level 7 (HL7) protocol has been designed to standardize the data transfer within hospitals. It is an application level protocol and so relates to level seven of the ISO/OSI-protocol hierarchy. HL7 covers various aspects of data

exchange in hospitals, e.g. admission, discharge and transfer of patients, as well as the exchange of analysis and treatment data. The HL7 standard represents hospital related transactions as standardized messages. HL7 is a de-facto standard for data exchange between commercial systems for hospitals.
More information is available at http://www.mcis.duke.edu/standards/HL7/hl7.htm

Manufacturing

OMG Manufacturing DTF Within the standardization efforts of the Object Management Group (OMG) Domain Task Forces (DTF) several Business Object Facilities are standardized. The Manufacturing DTF is one of them.
More information is available at http://www.omg.org/

STEP The STEP Standard for Product Data Exchange is a data transfer standard which supports design, reuse, and data retention, and provides access to data across a product's entire life cycle. STEP is an actual file format (i.e., a common neutral format for exchanging CAD/CAM data among dissimilar systems).
More information is available at http://www.ukcic.org/step.

Enterprise Application Integration

OAGIS The Open Applications Group is a non-profit industry consortium of business-application software vendors. The Open Applications Group Integration Specification (OAGIS) defines inter-application integration scenarios, and details the content required for connecting the applications. The OAGIS encompasses interoperability specifications for front-office, back-office, and supply-chain business-software components.
More information is available at http://www.openapplications.org/

OMG Within the standardization efforts of the Object Management Group (OMG) Domain-Specific Task Forces several Business Object Facilities are standardized. The scope of the OMG's Financial Domain Task Force (CORBAfinancials Task Force) comprises financial services and accounting. The Insurance Working Group is part of the OMG's Financial Domain Task Force. The Insurance Working Group aims at developing domain-specific interfaces that will enable insurance companies and other financial institutions to leverage purchased componentry and integrate their data. Another area is addressed by the Manufacturing Domain Task Force, whose goals are interoperable manufacturing domain software components through CORBA technology.
More information is available at http://www.omg.org/

Electronic Commerce

With respect to the situation in the area of electronic commerce, we can quote Andrew Tannenbaum:

"The good thing about standards is that there are so many to choose from."

Obviously, a consolidation is required in this domain. Enterprise application integration is highly related to electronic commerce applications.

OMG Electronic Commerce DTF Within the standardization efforts of the Object Management Group (OMG) Domain Task Forces (DTF) several Business Object Facilities are standardized. The Electronic Commerce DTF is one of them. More information is available at http://www.omg.org/

OBI The Open Buying on the Internet (OBI) Consortium is a nonprofit organization dedicated to developing open standards for business-to-business Internet commerce. The OBI Consortium is an independent collaborative managed by CommerceNet. More information is available at http://www.openbuy.org/

OFX Open Financial Exchange (OFX) is a specification for the electronic exchange of financial data between financial institutions, business and consumers via the Internet. More information is available at http://www.ofx.net/

OTP The Open Trading Protocol (OTP) was developed by a number of organizations, working cooperatively to make widespread Internet trading a convenient and secure reality. OTP is a protocol for the development of software products to permit product interoperability for the electronic purchase that is independent of the chosen payment mechanism. OTP encapsulates the payment with the offers/invoice/receipts for payment and delivery. More information is available at http://www.otp.org/

PIP RosettaNet is an international organization dedicated to the adoption and deployment of open and common business interfaces in the IT industry. It defines Partner Interface Processes (PIP) to provide common business/data models and documents enabling system developers to implement RosettaNet eBusiness interfaces. More information is available at http://www.rosettanet.org/

UN/EDIFACT This standard defines the United Nations rules for Electronic Data Interchange For Administration, Commerce and Transport, which comprises guidelines for the electronic interchange of structured data, and in particular that related to trade in goods and services between independent, computerized information systems. More information is available at http://www.unece.org/trade/untdid/Welcome.html

REFERENCES

Batini, C., Lenzerini, M., & Navathe, S. (1986). A Comparative Analysis of Methodologies for Database Schema Integration. *ACM Computing Surveys*, 18(4):323-364.

Booch, G., Rumbaugh, J., & Jacobson, I. (1999). *Unified Modeling Language User Guide*. Object Technology Series. Addison-Wesley, Reading, MA.

Conrad, S., Eaglestone, B., Hasselbring, W., Roantree, M., Saltor, F., Schönhoff, M., Strässler, M., & Vermeer, M. (1997). Research Issues in Federated Database Systems (Report of EFDBS '97 Workshop). *SIGMOD Record*, 26(4):54-56.

Dayal, U. & Hwang, H.-Y. (1984). View Definition and Generalization for Database Integration in a Multidatabase System. *IEEE Transactions on Software Engineering*, 10(6):628-644.

Dikel, D., Kane, D., Ornburn, S., Loftus, W., & Wilson, J. (1997). Applying software product-line architecture. *Communications of the ACM*, 30(8):49-55.

Elmagarmid, A., Rusinkiewicz, M., & Sheth, A., editors (1998). *Management of Heterogenous and Autonomous Database Systems*. Morgan Kaufmann.

Hurson, A. R., Bright, M. W., & Pakzad, S. (1993). *Multidatabase Systems: An Advanced Solution for Global Information Sharing*. IEEE Computer Society Press.

Jennings, N. & Wooldridge, M., editors (1998). *Agent Technology: Foundations, Applications, and Markets*. Springer-Verlag.

Kahng, J. & McLeod, D. (1998). Dynamic classificational ontologies: Mediation of information sharing in cooperative federated database systems. In Papazoglou, M. and Schlageter, G., editors, *Cooperative Information Systems: Trends and Directions*, pages 179-203. Academic Press, San Diego.

Kleewein, J. (1996). Practical issues with commercial use of federated databases. In *Proc. 22th International Conference on Very Large Data Bases (VLDB'96)*, page 580, Bombay, India. Morgan Kaufmann.

Missier, P., Rusinkiewicz, M., & Silberschatz, A. (1995). Providing multidatabase access - an association approach. In *Proc. Sixth Workshop on Database Interoperability*.

Paepcke, A., Chang, C.-C., Garcia-Molina, H., & Winograd, T. (1998). Interoperability for digital libraries worldwide. *Communications of the ACM*, 41(4):33-43.

Schatz, B. & Chen, H. (1996). Building large-scale digital libraries. *IEEE Computer*, 29(5):22-26.

Schmitt, I. & Saake, G. (1998). Merging inheritance hierarchies for database integration. In Halper, M., editor, *Proc. Third IFCIS International Conference on Cooperative Information Systems (CoopIS'98)*, pages 322-331, New York City, NY. IEEE Computer Society Press.

Schrefl, M. & Neuhold, E. (1988). Object Class Definition by Generalization Using Upward Inheritance. In *Proceedings 4th International Conference on Data Engineering (ICDE'88)*, pages 4-13. IEEE Computer Society Press.

Shaw, M. & Garlan, D. (1996). *Software architecture: perspectives on an emerging discipline*. Prentice Hall.

Sheth, A. (1998). Changing focus on interoperability in information systems: From system, syntax, structure to semantics. In Goodchild, M., Egenhofer, M., Fegeas, R., and Kottman, C., editors, *Interoperating Geographic Information Systems*. Kluwer.

Sheth, A. and Larson, J. (1990). Federated database systems for managing distributed, heterogeneous, and autonomous databases. *ACM Computing Surveys*, 22(3):183-236.

Taylor, R., Tracz, W., and Coglianese, L. (1995). Software development using domain-specific software architectures. *ACM SIGSOFT Software Engineering Notes*, 20(5):27-38.

Wiederhold, G., editor (1996). *Intelligent Integration of Information.* Kluwer Academic Publishers, Boston.

Chapter IX

Establishing Standards for Business Components

Klaus Turowski
Otto-von-Guericke-University Magdeburg

INTRODUCTION

The field of business applications encompassing software for production planning or order management is dominated by some big vendors offering large standardized off-the-shelf application systems, e.g. SAP R/3, BAAN IV, or Oracle Applications. Despite all advantages of selling and buying these large integrated application systems, there are two major shortcomings: Usually only large enterprises can afford to buy, install, customize, and maintain these systems, and the complexity which is inherent to these still growing systems makes it harder and harder for vendors to enlarge and maintain them. Furthermore, markets for these large packaged application systems are almost saturated, because most of the large enterprises have covered their demand. On the other side, there are lots of small and medium enterprises (SME), which demand very specialized application systems together with a broad coverage of *common* business functionality, as the core of the related business tasks stay the same—independent from a company's size. However, nearly no SME can afford to purchase or even to maintain these large packaged application systems. This leads to two important problems: SMEs demand on business application systems can not be satisfied, and vendors of large standardized off-the-shelf application systems lose an important market. In addition, small and medium vendors of business application systems, which offer highly specialized applications, have to master high barriers to entry the software market, as they have to offer core business functionality as well, in order to sell their specialized products.

A way to avoid this emerging software (industry) crisis in the area of business application systems, is to achieve a goal, which is followed up for a long time by the software engineering research community: to assemble software by using predefined reusable parts, so-called (software) *components*. Definitions of what exactly constitutes a component vary in the literature. For definitions and an overview of definitions given in the literature . (Szyperski, 1998, p 164-168). In common, software components are mostly understood as self-contained units with contractually specified interfaces and functionality. They may be sold independently, and be composed by third parties (e. g. to form new application systems) in combinations not assumed by the component's manufacturer.

Since we focus on the area of business application systems, we will use the term *business component* (BC) instead of component to emphasize our focal point. BC is a specialization of the term *component*, which names components that implement a set of functions out of a given *business* domain, e. g. finance or order management (Fellner, Rautenstrauch, & Turowski, 1999). For components in general, e. g. for graphical user interfaces or database access, some but not all of the assumptions made or conclusions drawn in the following may be valid.

Even though the combination of goals differ from each other according to a company's size or role, e. g. large vendor or small customer, important driving goals behind systematic software reuse based on component-orientation are expected improvements concerning cost efficiency, quality, productivity, market penetration, market share, performance, interoperability, reliability, or software complexity.

Depending on the strategy a company follows up with component-orientation, different kinds of standards develop in variant *standardization scenarios*, e. g. a large vendor of off-the-shelf application systems trying to reduce his software's complexity may define a *proprietary "standard"*, like Baan or SAP. On the other hand, *industry standards* may be (automatically) established by best practice, or *open standards* may be proposed by (independent) public organizations as the OMG (Object Management Group), not at last to expand markets and to avoid oligopolies. Keeping in mind that all mentioned scenarios have specific advantages for respective market participants, we will focus our contribution on discussing scenarios related to open standards, since it is common opinion that most improvements related to component-orientation will not prove true in this scenario if standardization fails.

After a short introduction to component-based business application systems, we address existing standardization approaches and the modeling layers to which they belong. E. g., some basic approaches, which belong to the technical layer are already available, like CORBA (Common Object Request Broker Architecture), COM (Component Object Model), JavaBeans, or mediators. However, no single approach is generally accepted. On the domain-related semantic layer, a much bigger lack of standardization can be observed. One reason for this lack of standardization is because there are too many "standards," and entities that try to establish their products as (industry) standards, often are just pretending to be component-oriented. Besides well-known approaches, e. g. from the OMG or OAG (Open Applications Group), contributions from standardization efforts in other areas are addressed, which do not originally focus on the task of standardizing BC, but are very well usable for this task, like UN/EDIFACT (United Nation's Electronic Data Interchange for Administration, Commerce and Transport), BSR (Basic Semantic Repository), or business *reference models* as proposed by the German Business Information Systems research community.

Another reason for the lack of standards for BC is, that there is still a lack of fundamental research work, concerning the question of the "right" usability and flexibility, or the "right" granularity of components from a technical and an economic viewpoint.

After examining *what* a standardized BC distinguishes from a non standardized one, looking at standardizing BC from a technical *and* an economic viewpoint will be the driving force of our reflections. We combine the technical and domain-related layers, which are both important for a successful standardization, by motivating the technical contents economically, and vice versa, by explaining the economical effects of technical decisions on standardizing BC.

This leads to the main part of our contribution: The analyses of shortcomings of existing standardization efforts for BC, the conclusions drawn to improve existing standards or to set up new ones, and the formulation of success factors for standardizing BC. Among other things we will discuss if it is too idealistic to demand *the* standard for BC or not, and how existing standards may be utilized. During this discussion, we address the importance of being able to solve content-related conflicts for a standard's acceptance by market participants, the role of a standard's extensibility and its realization, and the sole definition of core standards.

COMPONENT-BASED BUSINESS APPLICATION SYSTEMS

Application scenario – how BCs are supposed to be used

The basic idea behind BCs is to provide a straightforward way to build business application systems by choosing (or creating) BCs that provide the functionality needed. BCs that are usable in unpredictable system configurations are easy to integrate, flexible, and allow customers to tune their business application systems by buying the best of breed to meet their company's needs. (Standardized) BCs of different vendors can be plugged together to achieve an individual solution. Among other reuse techniques, e. g. generation or code and design scavenging (Sametinger, 1997, pp. 25-28), we focus on *compositional reuse*. We explain compositional reuse in the following example.

Suppose the scenario illustrated in figure 1: Three (software) companies (*vendor A*, *B*, and *C*) sell BCs. *A*, *B*, and *C* offer, in addition to other BCs, one or more BCs with similar functionality. In the example, two venders offer the BCs *purchase* and *financials*. A customer installing a new or changing an existing business application system can now choose the BC *purchase*, which supports his procurement business process at its best. *Customer 1* and *2* choose this BC from *vendor A*. *Customer 2*, who also operates an application system for production planning and control, decides to use additionally the BC *materials management* from *vendor A*. However, he could have chosen a BC from another vendor, since all BCs offered on the market, by definition, work together.

Furthermore, *customer 1* has to support the financial and the sales area with an (integrated) application system. As there is more than one vendor that offers a BC *financials*, he can compare the functionality of each and determine the one which meets the requirements of his business processes. Selecting the BC *sales* is carried

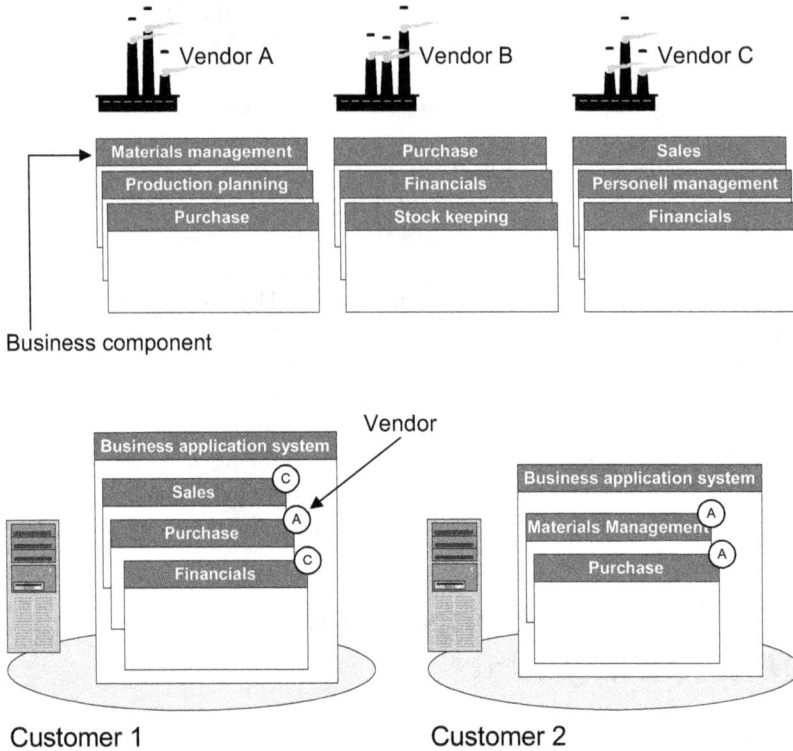

Figure 1: Building business application systems out of standardized BCs

out the same way. Although, we only show one vendor (*vendor C*) that offers an appropriate BC, there might be other venders that sell an equal BC as well. After choosing BCs, they are integrated in the respective application systems. By this, they add new functionality, or they replace older functionality. Nevertheless, replacing functionality may lead to *content-related (semantic) conflicts*, if more than one BCs offer the same functionality.

Suppose that *customer 2* would like to replace his stock keeping software. As depicted in figure 1 he can easily do this by buying the BC *stock keeping* offered by *vendor B*. However, this leads to a conflict since a BC *materials management* will mostly include this functionality already. We come back again to this point when we discuss success factors of standardizing BCs.

Besides BCs, an application system based on BCs needs additional software that do not belong to the core application itself, but do provide services needed from different application systems (cp. figure 2), e. g. user interfaces or print services, and software that supports a components life-cycle during development of the application system (Brown & Wallnau, 1996, pp. 8-12).

In addition, BCs contain exclusively domain-specific functions. Application systems made up of BCs need therefore an additional cross-domain component, which enables coordination at the technical *and* the business level. A *component application framework* encapsulates the services and tasks mentioned. A component application framework may even encompass major parts of the application itself, e. g. San Francisco from IBM (IBM, 1998).

Middleware offers cross-domain services, not supported by the respective operating system. We distinguish between operating system near and application near middleware services. Services close to the operating system manage technical communication issues between BCs like remote function calls (RFC). Application near middleware services can be classified as specific application software, like database management systems (DBMS) or software that enables integration of different BCs to a complex application like workflow management systems (WFMS). Furthermore, a *component system framework* is needed that supports composing and wiring of BCs, e. g. object request brokers. In general, middleware supports only technical aspects of collaboration, but not business domain-specific aspects.

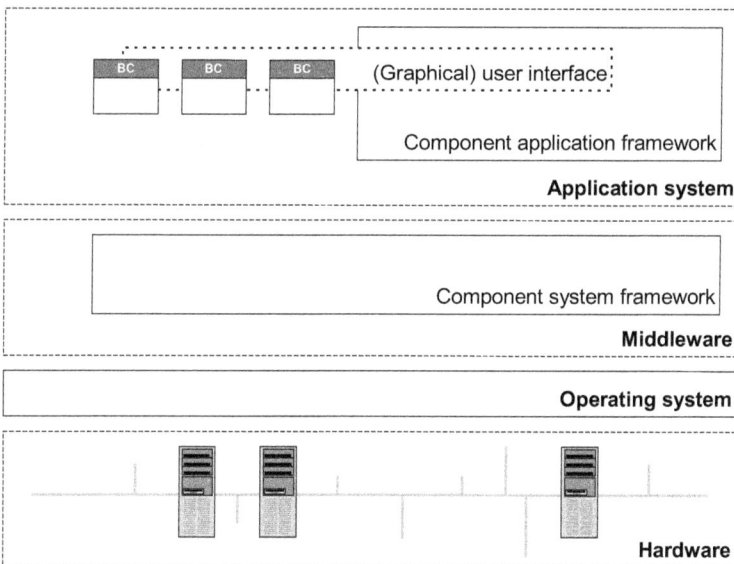

Figure 2: General architecture of component-based application systems

Driving goals of component-based software reuse

In the past, companies had to decide whether to make (individual software) *or* buy (standardized software) in order to get the required application system. Meanwhile standardized (off-the-shelf) business application systems offer open programmable interfaces, which allow an integration of existing individual software. The way to assemble an effective and efficient portfolio of business applications leads to a "make *and* buy" instead of a "make or buy" (Kurbel, Rautenstrauch, Opitz, & Scheuch, 1994):

- Standardized business processes are supported by standardized business application systems.

- Individual software implements business processes that help to improve a respective company's market position.

An efficient and effective make *and* buy is supported by component-based development of business application systems, which is based on (standardized) software components – the BCs. BCs may be parts of off-the-shelf business application systems, or may be implemented as individual software. Thus, BCs combine the advantages of individual software (e. g. low functional ballast, less training needed) with those of standardized software (e. g. short time to operation, low risk of failure, and more functionality).

Customers can buy the most suitable BC on the market or develop it independent from any other. The advantages for buyer's are obvious: The best available solution can be assembled vendor-independently (cp. figure 1). At the same time, the risky and long-term process of developing the needed software can be avoided. Since BCs can be modified and exchanged independently, the assembled application system is easily adaptable to specific business situations.

Buyer		Vendor	
Advantages	*Disadvantages*	*Advantages*	*Disadvantages*
Has to buy only necessary BCs	More complicated and expensive customizing	Smaller barriers to entry the BC market	Less dependence of buyers
Increased possibility to tune to his company	Additional effort for composition	Shorter time to market	Stronger competition
Can buy most outstanding BCs	Unclear responsibilities in composed systems	May address special market segments	Buyer's markets in widespread BCs
Make use of a vendor's principal competencies		Reduced development costs	Standardization effort
Less vendor dependence		Reduced maintenance costs	
Higher flexibility		Reduced complexity	
Better comparability		Better software quality	
Competition, possibly buyer's markets		Wider markets through platform independence	
Easy integration of legacy applications			

Table 1: Advantages and disadvantages of component-based business application systems

Customers knowing a huge number of vendors that offer a specific component can choose the appropriate BC based on a comparison process. This increases the competition between vendors, and will lead sooner or later to cheaper and more powerful BCs, thus providing better support for different business processes. However, customers buy BCs not solely based on their capability to support a specific business process. Soft facts like personal preferences to one or the other system or company, confidence in a company's wealth, former experiences with companies or their products, and involved costs may be considered when it comes up for a decision.

On the other hand, customizing an application system to a company's needs may be more complicated and more expensive if different vendors use different customizing techniques. Furthermore, additional effort may be necessary to compose an application system out of BCs, and it may be unclear which vendors are responsible if some parts of composed applications fail.

Vendors on the other hand are not forced by the component market to offer all possible BCs for an integrated application system. They can specialize on a certain area of business applications or offer BCs for a niche market, e. g. a stock keeping BC for the foundry market. Special market segments can be adressed more goal-directed by putting different BCs together. With this, barriers to enter the BC market are diminished and time to market is shortened. Because joining of BCs is performed through well-defined standardized interfaces, each component can be installed on different platforms—as long as the underlying middleware supports distributed components. This platform independence creates a wider market.

After all, costs for development and maintenance are reduced. However, on the long run, only those will survive, who find areas, where they have more competence compared to competitors, thus offering better BCs.

A summary of the discussion of component-based software reuse shows table 1, cp. e. g. (Orfali, Harkey, & Edwards, 1996, pp. 29-32), (Sametinger, 1997, pp. 11-19).

Why component-orientation fails without standardization

To use BCs as described in our application scenario, there have to be standards for BCs in each application domain, addressing the following:

- The component standard has to be aimed at the user's needs and should allow a short time-to-market of BCs based on already available middleware.
- Standardized BCs have to be independent at the technical level as well as at the business level to assure exchangeability of BCs from different vendors. Furthermore, minimum requirements of an open middleware, serving as a basis for the application system composed by BCs, have to be specified.
- BCs have to be standardized by means of a publicly available interface model.
- To support a specific business process it has to be possible to compose a number of fine-grained, standard-compliant BCs from eventually different sources.
- Any business application must be able to use required services from BCs out of the composed component-based business application systems.
- BCs have to cooperate at the business as well as the technical level. This allows

business driven changes to put into operation by setting necessary parameters (customizing) or exchanging whole BCs.

Since the standards mentioned are still missing in the area of business application systems, the advantages given in table 1 do not prove true. Besides a standardized middleware, which encompasses a distribution and integration model in the application domain (if BCs are supposed to work in "plug-and-play" mode (self-installing, self-testing)), the most important prerequisite is the standardization of application domain specific services and interfaces.

CLASSIFYING STANDARDIZATION APPROACHES FOR BCS

Figure 3 shows different entities that may be the objects of a standardization approach. On the top, there are business ontologies. A business ontology is a common term set for a specific area of concern. It provides basic definitions and explains the semantics of terms used, e. g. for production planning and control terms like *part* or *order* are defined in the corresponding business ontology.

Conceptual models are based on business ontologies. They describe business data, business functions, and business processes. In addition, they explain relations between their objects of concern according to certain views (e. g. data view, function view, or process view) and between objects of different views (e. g. which business functions operate on customer orders).

Since conceptual models mostly remain on a relatively high level of abstraction, a more detailed level of abstraction is necessary to complement the description of a given business domain. For this, attributes and parameters have to be taken into account. The definition of business data interchange formats and interfaces support this task. Business data interchange formats describe attributes of business data in detail, e. g. attributes that belong to a customer, or characters that are allowed to spell his name. However, these formats are mostly defined for communication purposes, and therefore, some information necessary to wholly describe a business domain may still be missing. Interfaces are mostly defined in a similar way as data interchange formats. Although, they describe parameters and return values needed for the invocation of services.

As depicted in figure 3, we have to consider all so far mentioned objects of standardization in order to standardize BCs. Standardization may occur on different modeling layers dependent on the objects of standardization. For our purposes, we distinguish between a *business domain-related* layer and a *technical* layer. The information needed to standardize BCs is mainly located on the business domain-related layer. However, boundaries between these layers are fluid.

Besides the pure standardization of BCs, it is necessary to standardize a component application framework together with a component system framework.

Another characteristic to classify standardization is to look at standardization scenarios. Standardization scenarios are determined through the subject that presses ahead with standardization and its intention. As an example, there are several advantages if solely one entity tries to establish a standard. In this case, we speak from a *proprietary standard*. Although a proprietary standard may be faster to

Domain-related

| Business ontologies |
| Conceptual models for business data, functions, and processes |
| Business data interchange formats |
| Interfaces |

Necessary for definition of BCs

Necessary for using application systems based on BCs

Layer

| Component application framework |
| Component system framework |

| Middleware |

| Basic techniques like XML |

Technical

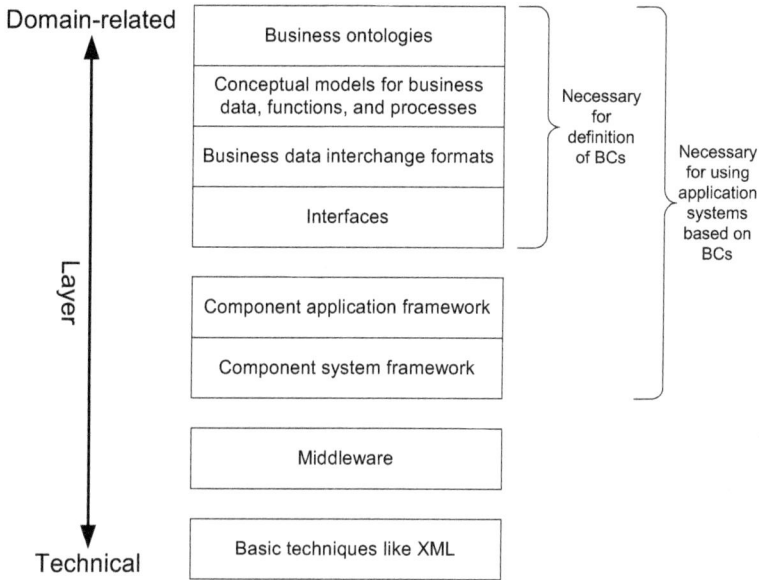

Figure 3: Classification of objects of standardization

establish, the main (mostly economic) advantages may lie by its originator (cp. the SAP example given in the next section). Thus, the standard leaves only a lower acceptance in the independent part of the market.

An *industry standard* is set up if most vendors base their products on an architecture of a successful competitor, e. g. the personal computer (PC) invented by IBM.

We will speak of an *open standard*, if an independent organization or an organization that encompasses a large group of major market participants creates the standard, e. g. the OMG or the United Nations.

In the following, we discuss existing standardization approaches. This discussion will provide arguments needed for the further analysis of shortcomings of existing standardization approaches for BCs and the formulation of success factors for standardizing BCs.

EXEMPLARY STANDARDIZATION APPROACHES

Business data interchange formats, conceptual models, and business ontologies

Business domain-related approaches mostly refer to support *electronic data interchange* (EDI) between different companies, or to approaches, which improve standards for EDI. EDI aims on organizational information-based surplus values (cp. Kuhlen, 1996, esp. p. 90), e. g. an improved organizational and operational structure as well as time and cost savings. The most important standard for cross-organizational data interchange was established by the United Nations with UN/ EDIFACT (electronic data interchange for administration, commerce, and transport) (UN, 1995). It standardizes electronic exchange of structured information, e. g. orders or invoices, thus permitting a direct communication between different business application systems. Due to fundamental drawbacks, like the missing of semantic rules, e. g. for quantity or packaging units, the implicit assumption that each organization uses similar business processes and scenarios (cp. Zbornik, 1996, pp. 92-93), and economic (e. g. high implementation costs) as well as organizational (e. g. slow adoption to changing business processes, complicate adjustment of established business process and rules) (cp. Goldfarb & Prescod, 1998, pp. 106-110) causes, that UN/EDIFACT did not win the expected recognition and implementation extent.

Open-EDI/object oriented-EDI (TMWG, 1998), Universal Data Element Framework (UDEF) (Harvey et al., 1998, pp. 25-26), Basic Semantic Repository (BSR), and its successor BEACON (Steel, 1997) are efforts that address the problems mentioned first. They focus on establishing uniform business scenarios and semantic rules. Furthermore, they aim on establishing *business ontologies*. Backward compatibility is supported by repositories, which allow to store relations between known standards, especially UN/EDIFACT, and the newly created ones (cp. e. g. Harvey et al., 1998, pp. 10-12). Other projects that deal with ontological problems concerning component-oriented development of applications are enumerated in (Szyperski, 1998, p. 318). Although these projects focus on the problem of how to name and find the components needed, they still contribute to the solution of providing content-related descriptions of BCs.

The XML/EDI-Initiative (Peat & Webber, 1997) on the other hand concentrates on economical and organizational drawbacks of UN/EDIFACT using the extensible markup language (XML) (cp. Bray, Paoli, & Sperberg-McQueen, 1997) to lower implementation costs and increase flexibility of EDI. The basic idea of the XML/EDI-Initiative, which is to encapsulate business data between XML tags, will be taken up later in the context of extending BCs.

The information systems research community also does a lot of work in the research field of *reference models*. Reference models give recommendations for the design of information systems at the business domain-related layer. Conceptual models may be found in Mertens & Griese (1993), Scheer (1994), Becker and Schütte (1996), or Mertens and Griese (1997). These contributions can be classified as preliminary, but nevertheless necessary for component-oriented software development because reference models classify and arrange tasks at the business level without defining any software components or proposing how to derive them.

Further work in this context is given in (Buxmann, 1996), where basic approaches to an optimal solution for standardizing interfaces to business application systems are developed. A domain-related object-oriented approach, which may serve as a proposal to standardize materials management, is given in (Rohloff, 1995. pp. 245-289). (Kemmner, 1991) and (Turowski, 1997, pp. 175-221) provide (object-oriented) conceptual models to a distributed production planning and control, which may be used as a basis to find BCs.

Business data interchange formats, ontologies, and conceptual models are suitable for further use in standardization approaches as depicted in figure 3. They provide basic knowledge of the respective business domain. However, they do not solve the problem of how to find BCs. A problem that is strongly related to the problem of how to find objects (classes) (cp. e. g. Lehner & Sikora, 1994, p. 40).

Proprietary and industry standards

SAP, as an example of an actual vendor of business application systems defines their own *business objects* (SAP, 1998). To open the system for legacy and custom-made applications these SAP business objects are callable from outside via the *Business Application Program Interface* (BAPI) (SAP, 1997).

SAP defines a proprietary interface standard to their application system R/3 by redefining the existing data and function model to an object and component model. The resulting object model is fine-grained and "reusable" on every platform SAP's R/3 is available. SAP business objects are coordinated at the business level by the application layer of the R/3 system and can be adapted at run-time by changing parameters (customization). New objects and components can be implemented using third-party development tools or the integrated *Advanced Business Application Language* (ABAP/4).

Since SAP business objects and BAPIs that belong to them are strongly system dependent and may not be used in unpredictable system configurations, the SAP approach does not follow the main idea that stands behind the definition of BCs. It is rather a proprietary interface to SAP's system R/3

The framework of predefined classes for business applications from IBM, the *San Francisco project* (cp. e. g. IBM, 1998), is based on JavaBeans technology from Sun Microsystems (Sun Microsystems, 1997). Business requirements that are not supported by the actual available system may be implemented by exchanging or adding parts of the class hierarchy. Contrary to our definition of BCs, the IBM framework allows direct manipulation of contained classes. This may lead to different implementations of a given framework version, cutting down the possible reuse of frameworks. For this reason, the IBM framework does not standardize BCs. However, classes of the framework, that in the future turn out to be stable, may be used as a basis for standardization.

"Open" standards

The OMG proposes to standardize selected business areas (domain services), e. g. manufacturing or financial, as a part of their Common Object Request Broker Architecture (CORBA). However, the standardization of the required underlying component model has not been adopted yet (OMG, 1997, p. 19).

Furthermore, the OMG component model focuses on *technical* cooperation based on the CORBA standard rather than on the cooperation at the semantic level, which is necessary for collaborating BCs. However, the technical communication layer is important, too. As an example, it is used to publish services of a BC and call services from other BCs.

Another necessary prerequisite for the OMG business component model is the *business object*. Business objects are real-world entities that belong to a given (business) field (OMG, 1996a, p. 19). For example, objects like *customer*, *car*, or *invoice* are defined as business objects. BCs manipulate these business objects.

Based on the CORBA standard the OMG standardizes business objects and a component model. The component model of the OMG standardizes coordination and cooperation between components at the technical level. The OMG's definition of business objects is the most accurate and can be found in many other publications. The component model of the OMG will be build around that object definition for distributed components, their coordination, and cooperation. The objects, respectively the components, can be reused domain-independently, on any platform where an object broker is available and extended by adding or changing objects.

The Open Application Group (OAG) defines *Business Object Documents* (BOD). BODs contain syntax and semantics of the information to be exchanged between coarse-grained business functions (OAG, 1997), e. g. manufacturing, financials, or human recourses. Although the business functions, which may be treated as BCs, are coarse-grained, the well-defined BODs may aid to describe and standardize fine-grained BCs.

Standards for the technical layer

Standards for the technical layer mostly concern middleware and the component system framework but also approaches to define interfaces of BCs technically.

The use of "off-the-shelf" BCs requires uniquely identifiable BCs, which can be easily integrated in existing systems. Therefore, the wrapped system as well as the BCs interfaces must be described in the same system independent definition language (e. g. the Interface Definition Language (IDL) suggested by the OMG (OMG, 1996b) or a Web IDL as proposed in (Merrick & Allen, 1997)).

Besides defining interfaces, there are several different approaches to provide the middleware needed to compose BCs, e. g.:

- OMG's Object Management Architecture (OMA) together with CORBA,
- Sun's Java and JavaBeans, or
- Microsoft's Component Object Model (COM) together with ActiveX.

In principle, all mentioned approaches are suitable to support business application systems that are based on BCs by providing a component system framework. Since we focus on the standardization of BCs, we refer to the literature for an in deep description and discussion of the various approaches, cp. e. g. (Orfali et al., 1996, pp. 43-532), (Szyperski, 1998, pp. 169-258).

Additional middleware products that can be used to complement business application systems based on BCs, like WFMSs and their interfaces are too the object of standardization. As an example the Workflow Management Coalition (WFMC) standardizes interfaces to allow the support of inter-company workflows (Hollingsworth, 1995).

MAJOR PROBLEMS OF STANDARDIZING BCS

Before looking at particular problems, we have to address standardization in general, by pointing out, that it seems to be too idealistic to demand *the* standard for BCs, which is suitable for any current and future business scenario. Final standards are ideals, at least in the area of business applications. With this, we follow the common thesis that our area of consideration changes faster than a complete standardization of contained BCs could occur.

However, even a standard that covers only a part of a business domain may be a real advance, as

- advantages of component-based business applications (cp. table 1) appear,
- a successful applied standard (even for a part of a certain business domain) may lead to a further standardization in order to achieve same advantages in other areas, and
- further functionality needed is still available by integrating non component-based applications.

One reason for standardization problems stems from the fact, that there are too much "standards" and entities that try to establish their products as (industry) standards (cp. section 4). As these entities mostly pursue different goals within their standardization approach, which are often conflicting, there is no common goal. Even definitions and basic concepts defer from each other, e. g. concerning type of reuse (black box, white box), time of reuse (during software development, during composition of components), or the objects of reuse (patterns, business objects, BCs). In addition, we can observe in this context that standardization approaches start from scratch, without considering existing efforts, e. g. business ontologies, reference models, or business data interchange formats, which may contribute to the intended standardization approach.

Table 2 shows a summary of major problems of standardizing BCs. The problems are organized according to the layer they belong to.

Despite of advances with vertical reuse in some areas, e. g. frameworks that assist the development of graphical user interfaces (Krasner & Pope, 1988), there is still a lack of fundamental research work concerning the question of the "right" usability and flexibility, or the "right" granularity of BC from a technical and an economic viewpoint. This comes together with the open problem of how to solve

General	Defining final standards for BC is an ideal that can not be reached, at least in the area of business applications
	Conflicting goals of participants in standardization of BCs
Business layer	Too much standards and standardizing entities
	Lack of proper definitions and common views
	Unsolved problems concerning flexibility, granularity, and solving of content-related conflicts
	Little consideration of existing standards (in related areas)
	Little consideration of technical and market aspects
Technical layer	Imbalance between technical and domain-related approaches
	Little consideration of domain-related and market aspects

Table 2: Major problems of standardizing BCs

content-related conflicts, which can not solely be solved technically, but have to be solved on a domain-related (business) layer.

Content-related conflicts appear, if more than one BCs offer a certain service. If this happens, precautions have to be taken to avoid these conflicts. As these precautions affect not only questions that are related to a certain domain, but do also have technical aspects, technical and content-related viewpoints have to be considered. Besides, economic reflections are necessary to ensure that there is a market for the newly standardized BCs.

Looking on the technical side reveals that there is an imbalance between technical approaches concerning component platforms or wiring standards and domain-related approaches, which concern the definition of BCs. Technical approaches are well established compared to domain-related approaches, as a lot of research work has already been done in this area. Even middleware products like object request brokers are already available. Although no approach is generally accepted, there are still market shares big enough to justify considering them. With this, known middleware products are principally suitable for supporting component-based business applications. However, technical approaches often suffer from not considering domain-related problems or market aspects.

LESSONS LEARNED AND CONCLUSIONS FOR FUTURE STANDARDIZATION APPROACHES

The analyses of shortcomings of existing standardization efforts together with the discussion of major problems of standardizing BCs given above, leads us to the formulation of success factors for standardizing BCs. Table 3 shows a summary of them.

In the following, we discuss the success factors separately, and show how they influence the improvement of existing standards or the set up of new ones.

During this discussion, we especially address the importance of being able to solve content-related conflicts for a standard's acceptance by market participants, the role of a standard's extensibility and its realization, and the sole definition of

---

Simultaneous consideration of technical, domain-related, and market aspects while standardizing BCs

Standards that allow the intersection of BCs offered by different vendors: the right granularity

Fine-grained standards

Foundation on and integration of existing standards while creating new ones, or at least providing references

Definition of core standards

Providing Extensibility

Providing means to solve content-related conflicts

Integration of legacy applications

Providing open standards

Table 3: Success factors for standardizing BCs

core standards.

Combined view – considering techniques, domains, and markets

After all, component orientation is economically motivated, e. g. reducing development costs of business application systems. Furthermore, BCs have to be sold, or at least software that is made of BCs. This leads to the demand that the combination of a general architecture of a component-based business application system, the BCs the application system is made of, and the domain specific standard to which the BCs refer has to be a marketable entity. Especially if we focus on the value added by component-orientation: the opportunity to sell software in parts, besides other advantages that refer to improving software development.

On the other hand, specific demands force the use of certain technologies and architectures, e. g. if legacy applications have to be integrated. Even the question of *what* exactly to standardize can be influenced by this, as we will see later in the discussion of how to solve content-related conflicts and the role of standards by this task.

Another aspect that influences standardization is how marketable BCs differ from each other. The question is if any two BCs offered on the market have to cover either a totally different functionality or the same functionality, e. g. BCs for materials management or sales, but not BCs that provide partial functionality only like stock keeping, which normally is a part of materials management. With respect to markets forcing vendors to do so is to restrictive, as e. g. a big software vendor may offer a BC for materials management that covers the requirements of most companies, but a small vendor, e. g. specialized in specific requirements of process industry, may offer a BC for stock keeping, which fits very well to specific requirements of process industry. In addition, if standardization of BCs is not done at a low level, standardization on higher levels either leads to standardization of any offered BC (from different vendors), or restricts vendors in their implementation (e. g. only vendors who can effort the implementation of a whole BC for materials management could offer BCs in this area).

This aspect is strongly related to the question of the right granularity of BCs. Thus urging us to demand the standardization of relatively fine-grained BCs in order to allow a high flexibility of vendors with respect to the functionality included in a certain BC they offer.

In the following, we will return to the market aspects raised here by looking at standardizing BC from a technical *and* an economic viewpoint. We combine the technical and domain-related layers, which are both important for a successful standardization, by motivating the technical contents economically, and vice versa, by explaining the economical effects of technical decisions on standardizing BC.

Consideration of existing standards and integration of legacy applications

As shown previously, there are already many standardization efforts, which are in use in some areas and which represent high investments. Looking closer to these standards shows that there is no need to reinvent the wheel since parts of them can be reused for creating new standards. Conceptual models like reference models for specific business domains as production planing and control (e. g. Scheer, 1994) describe most functionality and most data needed in a given area. However, they are often restricted to a relatively high level of abstraction. Thus providing only core information needed for standardization. At this point considering complementary standards can be useful.

The data model given in Scheer (1994) describes the data needed only at the level of entities and relationships, but not at the attribute level. On the other hand, UN/EDIFACT describes messages and message formats to exchange business data, by this focussing on the attribute level. Although UN/EDIFACT does not cover all attributes relevant for the area of production planning and control, there is still a large intersection, thus complementing some of the information missing in the conceptual model.

Furthermore, backward compatibility can be provided, e. g. XML-EDI is a heading under which several new standardization efforts to exchange business data between companies may be summarized. Examples are provided in (CEN/ISSS, 1998). Due to the foundation on UN/EDIFACT backward compatibility is provided and older applications can be used further (Turowski, 1999). However, new requirements can be satisfied as well.

Backward compatibility of standards comes together with compatibility of legacy applications that are based on (legacy) standards. Therefore, backward compatibility of standards allows a smooth transition from old to new standards.

In general, any legacy application that is important for a certain company has to integratable in a new business application system that is based on newly standardized BCs. To support this, we have to provide technical means that are part of the basic architecture, which a certain standard needs, e. g. object request brokers together with IDLs and wrappers. Since legacy systems are mostly replaced step by step with new application systems, especially with those that are based on BCs, as in the beginning most of the BCs needed might not yet exist, integration of legacy applications and consideration of necessary integration during standartization is a critical success factor.

Besides better backward compatibility, existing standards (that are in use) are often accepted and a new standardization approach, which sets up on existing standards, or at least provides cross-references, may lead to a higher acceptance of the new standard. In addition, reuse of existing standards may accelerate the standardization process itself, leading to higher flexibility, better acceptance since requirements can be satisfied more timely, and not at last to reduced costs of standardization.

Solving content-related conflicts

As already discussed BCs may overlap one another and BCs may be built by combining other BCs. That means that different BCs may support the same services. For this reason, we have to solve content-related conflicts (cp. section 2) in order to allow the composition of intersecting BCs that might be sold and produced by different vendors. This is a problem where market aspects force specific technical solutions, and a certain approach to standardization.

By trying to determine the smallest reasonable intersections between any two BCs, one can find *elemental* BCs, and these can be used to solve content-related conflicts (Fellner et al., 1999).

As a sidenote, it is to mention that these intersection may be hard to find since an elemental BC should be big enough to be sold individually, but small enough to forbid further differentiation with respect to demand. On the other hand, an elemental BC must be big enough to justify independent development, maintenance, and sale.

As a result, marketable BCs should (virtually) be made of a set of standardized elemental BCs. Hereby, an important point is that there is still a high flexibility for vendors to create marketable BCs, as long as specific elemental BCs contained in a marketable BC can be turned on or off.

With respect to standardization, the task is left to standardize relatively fine-grained elemental BCs. We too have identified this task in section 6.1.

Sole definition of core standards

We have shown that defining final standards for BCs is an ideal, which will never be reached due to constantly changing business scenarios and because of the pure extent of (slightly) different business scenarios. However, we have also explained that building business applications by composing BCs do not work without standardizing BCs.

A possible solution of this problem is to define core standards with a built-in *extensibility*. Taking as an example an (elemental) BC that supports the storage of customer data. This BC must most probably provide attributes like a customer's name, his address, or communication data as telephone number, Web site, or e-mail address. However, 20 years ago web side or e-mail address would not have been important, but the customer as an entity with common attributes like name, address, or communication data, as well as its relationships to orders, accounts, or customer-specific terms and conditions would already have been known. Thus allowing all (legacy) BCs, which do not need the newly added attributes, to use the services originally provided unchanged. Nevertheless, new BCs, e. g. those that support EDI, may need these newly added attributes. At this point, the original standard has to be extended. The next section addresses approaches to support extensibility.

The conclusion drawn for standardization of BCs is that (together with extensibility) the definition of core standards is sufficient as a starting point. By this, standardization can be further accelerated and time-to-market can be shortened. Thus resulting in early offers of BCs, which may create further demand. However, extensibility has to be technically supported as well.

Extensibility

Standards and products (BCs) that rely on these standards have to be maintained. If they are not maintained, they will become unsaleable like most other products. Therefore, it is not sufficient to place once developed BCs in repositories. With this, another general conclusion, which we can draw, is that standards, especially for the dynamic area of business domains, have to be an object of continuous improvement. Thus making each standard for BCs a core standard that is subject to changes, especially extensions. In fact, it would be problematic to leave existing services out in future standards, since backward compatibility would suffer from this. For this reason, we focus on extensibility. In addition, empirical studies support the idea of extensibility, as customers prefer to reuse flexible components (Lerch, Flor, Fichman, & Hong, 1998).

Extensibility is the possibility to add any kind of services to an existing standard for BCs. New services can provide additional functionality, e. g. a new method for machine loading, or access to additional data, e. g. new communication data. While it is easy to add totally new functionality, it can be hard to integrate additional data, since certain anomalies may occur, e. g. update, insertion, or deletion anomaly (cp. e. g. Schlageter & Stucky, 1983, p. 164), or interfaces needed are missing.

Using new storage techniques, which combine storage of data and its description, may solve some of the arising problems. Assume it uses an XML tag with a standardized name to explain the semantics of what is stored. With this, interfaces can be simplified and additional data that may not be understood by older BCs can easily omitted.

The naming of the XML tags is once more a standardization task. However, approaches that focus on standardizing these names do already exist (cp. section 4.1). Furthermore, this constitutes another opportuinty to integrate and reuse existing standardization efforts in the standardization of BCs.

Other success factors

Another success factor is given with the *openness* of a standard for BCs. Openness means that BCs, which are based on an open standard, can be used together with different middleware products. Due to the fact that there are different component system frameworks, e. g. CORBA, COM, or JavaBeans, which do not per se cooperate with each other, this is a problem of abstraction, which primarily concerns the technical layer. Typical solutions for this problem are to use bridging products or additional layers that abstract from specific needs of a certain middleware product. However, during standardization it has to be ensured that there are no dependencies to middleware products, which lie underneath the additional abstraction layer. If standardization is done in this way, standards will be still usable, independent from a specific technical infrastructure. An example for this is UN/ EDIFACT, which serves well as a core standard and that experiences a major comeback due to new XML-based techniques.

Besides creating potentially marketable BCs that are based on a given standard, it is necessary to provide means that allow customers to find the BCs they need. This is too a problem of standardization. It concerns the proper naming of BCs and the proper description of what they do. This problem is also known from the area of manufacturing. There classification schemas together with group technology are used to overcome this problem. Similar approaches for our area of concern are discussed in (Szyperski, 1998, pp. 317-318).

OUTLOOK AND CONCLUSIONS

If we believe in market forecasts, there will be a high demand for component-based business application systems (Szyperski, 1998, pp. 18-20), not at last to avoid the emerging software industry crises. However, the main task is still open: The standardization of BCs.

Although the future may be bad, with respect to the named major problems, there is still a light at the end of the tunnel. Our main asset, as well as a large problem, may be the high number of standardization efforts and outcomes that come along with them. Their combination may lead to new standards that encompass older ones. Means to support this are repositories, cross references, and new techniques (e. g. XML) that allow combining data and the description of its semantic.

Besides analyzing shortcomings of existing standardization efforts for BCs, we have provided success factors for standardizing BCs. One of the main success factors for standardizing BCs is to consider techniques, domains, and markets altogether, instead of focusing on one single view. Furthermore, the consideration of existing standards as well as considering possible integration of legacy systems already during standardization is crucial. In addition, content-related conflicts have to be solvable, which as well influences standardization. At last, we have addressed the importance of the sole definition of core standards together with a standard's extensibility. Especially if we look on existing standards like UN/EDIFACT, this last point may become a key success factor, as business processes and scenarios are highly dynamic and the object of continuous improvement.

REFERENCES

Becker, J., & Schütte, R. (1996). *Handelsinformationssysteme*. Landsberg.

Bray, T., Paoli, J., & Sperberg-McQueen, C. M. (1997). Extensible Markup Language (XML) : http://www.w3.org/TR/PR-xml.html.

Brown, A. W., & Wallnau, K. C. (1996). Engineering of Component-Based Systems. In A. W. Brown (Ed.), *Component-Based Software Engineering: Selected Papers from the Software Engineering Institute* (pp. 7-15). Los Alamitos, California: IEEE Computer Society Press.

Buxmann, P. (1996). *Standardisierung betrieblicher Informationssysteme*. Wiesbaden: Deutscher Universitäts-Verlag.

CEN/ISSS. (1998). Interim Report for CEN/ISSS XML/EDI Pilot Project : http://www.cenorm.be/isss/workshop/ec/xmledi/interim.html.

Fellner, K., Rautenstrauch, C., & Turowski, K. (1999). Fachkomponenten zur Gestaltung betrieblicher Anwendungssysteme. *IM Information Management & Consulting, 14*(2).

Goldfarb, C. F., & Prescod, P. (1998). *The XML Handbook*. Upper Saddle River: Prentice-Hall.

Harvey, B., Hill, D., Schuldt, R., Bryan, M., Thayer, W., Raman, D., & Webber, D. (1998). Position Statement on Global Repositories for XML . ftp://www.eccnet.com/pub/xmledi/repos710.zip.

Hollingsworth, D. (1995). *Workflow Management Coalition: The Workflow Reference Model*. Winchester.

IBM. (1998). San Francisco Project Technical Summary : http://www.ibm.com/Java/Sanfrancisco/prd_summary.html.

Kemmner, G.-A. (1991). *Entwicklung eines Leitfadens zur betriebsindividuellen Teildezentralisierung EDV-gestützter PPS-Systeme. Schlußbericht zum*

Forschungsvorhaben Nr. 7574. Aachen: fir Aachen.

Krasner, G. E., & Pope, S. T. (1988). A Cookbook for Using the Model-View-Controller User Interface Paradigm in Smalltalk-80. *Journal of Object-Orientated Programming, 1*(3), 26-49.

Kuhlen, R. (1996). *Informationsmarkt: Chancen und Risiken der Kommerzialisierung von Wissen.* (2 ed.). Konstanz: Universitätsverlag Konstanz.

Kurbel, K., Rautenstrauch, C., Opitz, B., & Scheuch, R. (1994). From »Make or Buy« to »Make and Buy«: Tailoring Information Systems Through Integration Engineering. *Journal of Database Management, 5*(1994), 18-30.

Lehner, F., & Sikora, H. (1994). Ergebnisse einer Untersuchung über Objektorientierte Softwareentwicklung. *Information Management, 9*(1), 36-45.

Lerch, J. F., Flor, N. V., Fichman, M., & Hong, S.-J. (1998). Software reuse and competition: Consumer preferences in a software component market. *Annals of Software Engineering*(5), 53-83.

Merrick, P., & Allen, C. (1997). Web Interface Definition Language (WIDL) . http://www.w3.org/TR/NOTE-widl-970922.

Mertens, P., & Griese, J. (1993). *Planungs- und Kontrollsysteme in der Industrie.* (7 ed.). (Vol. 2). Wiesbaden: Gabler.

Mertens, P., & Griese, J. (1997). *Administrations- und Dispositionssysteme in der Industrie.* (11 ed.). (Vol. 1). Wiesbaden: Gabler.

OAG. (1997). Open Applications Integration : ftp://ftp.openapplications.org/openapplications.org/whtpaper.zip.

OMG. (1996a). Common Facilities RFP-4: Common Business Objects and Business Object Facility : ftp://ftp.omg.org/pub/docs/cf/96-01-04.pdf.

OMG. (1996b). IDL to JAVA RFP : ftp://ftp.omg.org/pub/docs/orbos/96-08-01.pdf.

OMG. (1997). CORBA Component Imperatives : ftp://ftp.omg.org/pub/docs/orbos/97-05-25.pdf.

Orfali, R., Harkey, D., & Edwards, J. (1996). *The Essential Distributed Objects Survival Guide.* New York: John Wiley & Sons.

Peat, B., & Webber, D. (1997). Introducing XML/EDI: "The E-business Framework" . http://www.geocities.com/WallStreet/Floor/5815/start.htm.

Rohloff, M. (1995). *Produktionsmanagement in modularen Organisationsstrukturen: Reorganisation der Produktion und Objektorientierte Informationssysteme für verteilte Planungssegmente.* München: R. Oldenbourg Verlag.

Sametinger, J. (1997). *Software Engineering with reusable components.* Berlin: Springer.

SAP. (1997). BAPI Catalog : http://www.sap.com/bfw/interf/bapis/preview/catalog/index.htm.

SAP. (1998). SAP Business Objects : http://www.sap.com/bfw/interf/objects.htm.

Scheer, A.-W. (1994). *Business Process Engineering: Reference Models for Industrial Enterprises.* (2 ed.). Berlin: Springer.

Schlageter, G., & Stucky, W. (1983). *Datenbanksysteme: Konzepte und Modelle.* (2 ed.). Stuttgart: Teubner.

Steel, K. (1997). The Beacon User's Guide: Open Standards for Business Systems . http://www.cs.mu.oz.au/research/icaris/beaug1.doc.

Sun Microsystems (Ed.). (1997). *JavaBeans: JavaBeans API Specification 1.01.* Mountain View: Sun Microsystems.

Szyperski, C. (1998). *Component Software: Beyond Object-Oriented Programming*. (2 ed.). Harlow: Addison-Wesley.

TMWG. (1998). Reference Guide: "The Next Generation of UN/EDIFACT": An Open- EDI Approach Using UML Models & OOT (Revision 12) : http:// www.harbinger.com/resource/klaus/tmwg/TM010R1.PDF.

Turowski, K. (1997). *Flexible Verteilung von PPS-Systemen - Methodik Planungsobjektbasierter Softwareentwicklung*. Wiesbaden: Deutscher Universitäts-Verlag.

Turowski, K. (1999). Agenten-gestützte Informationslogistik für Mass Customization. In H. Kopfer & C. Bierwirth (Eds.), *Logistik Management - Intelligente I+K Technologien* . Berlin: Springer.

UN. (1995). United Nations Directiories for Electronic Data Interchange for Administration, Commerce and Transport : http://www.unece.org/trade/untdid/ Welcome.html.

Zbornik, S. (1996). *Elektronische Märkte, elektronische Hierarchien und elektronische Netzwerke: Koordination des wirtschaftlichen Leistungsaustausches durch Mehrwertdienste auf der Basis von EDI und offenen Kommunikationssystemen, diskutiert am Beispiel der Elektronikindustrie*. Konstanz: Universitätsverlag Konstanz.

Chapter X

A Standards-Based Common Operational Environment

Jaroslav Blaha
NATO ACCS Management Agency, Belgium[1]

INTRODUCTION - THE NEED FOR A COE

NATO as an international organization consists of representatives from 19 nations and has to satisfy the political, operational and technical requirements of all members. In addition to the development of specific NATO information systems, it has to consider and support means for interoperability with national systems. Taking into account the life-cycle-cost aspects, it becomes clear that a standardization policy must consider the cost-effective maintenance, upgrade and replacement of system components and their interfaces.

Standardization of NATO specific interfaces is a well-established process, which results in 'Standardization Agreements' (STANAGs). Those are specifications of proprietary standards or of adaptations of international (e.g. ITU, ISO) standards. STANAGs have two major disadvantages: First, the process to develop a new or to modify an existing standard is lengthy; the process to get the specification ratified by the relevant nations is even longer. This can result in STANAGs, which do not reflect the state-of-the-art of standards. Second, as STANAGs often specify standards that are different from international or commercial standards, there is very little market and product support. This leads naturally to increased development and procurement costs.

In military terms a Command, Control and Information System (CCIS) is the equivalent to a Management Support and Information System (MSS/MIS) in the commercial domain.

The COE efforts originated with the simple observation that in command, control and information systems certain functions (e.g. message exchange, tasking, communication interfaces) are so fundamental that they are required for virtually every CCIS. Yet these functions are built over and over again, in often-incompatible ways, even for systems with almost identical requirements. If such common functions could be extracted, implemented as a set of extensible building blocks, and made readily available to system designers, development schedules could be accelerated and substantial savings could be achieved (although the quantitative demonstration of cost effectiveness is a complex problem). Moreover, interoperability would be significantly improved because common software is used across systems and the functional capability only needs to be built correctly once rather than for each project (DII-COE, 1999).

The COE currently focuses on CCISs and mostly ignores special-to-purpose systems outside this domain (e.g. satellite control, onboard navigation systems). CCISs are currently the most common and therefore preferred domain. However, it is expected that the COE concept will over time be extended into non-CCIS areas. One example is the Domain Specific Software Architectures (DSSA) project of ARPA, which attempts to standardize real-time avionics architectures through an Avionics Domain Application Generation Environment (ADAGE) (ADAGE, 1999).

Throughout this paper the following terms will be used:
- Common applications, modules or subsystems provide functionality, which is available and needed by effectively all users, independent of their specific task in the organization. Examples are e-mail or word-processing.
- Functional Area Sub-Systems (FASS) and associated applications provide functionality, which is required only by a limited group of users specifically for their tasks. Examples are logistics or human-resource management.

An integrative approach is needed, which could provide guidance to project managers, engineers and budget authorities on how to
- derive generic technical requirements from existing operational require-ments, i.e. by mapping of existing building blocks of technical solutions (e.g. an e-mail system) against a common class of operational needs (e.g. commu-nicate tasks effectively with remote subordinates),
- select state-of-the-art standards, which support the desired functionality, and are implemented in commercially available and market-proven products to avoid bespoke developments,
- aggregate standards and the associated products into building blocks with well-known and tested characteristics, which can be assembled to operational systems with minimal development effort, and
- to reduce system development risks, and to ensure the system's ability to evolve in order to benefit from information technology advances.

In addition this approach had to be harmonized with ongoing standardization activities, both on policy level (interoperability policies) and technical level (STANAG development). To allow smooth cooperation with the member nation's systems, the resulting NATO COE should also be harmonized and compatible with existing national COEs.

BACKGROUND - WHAT IS A COE?

The COE concept is an approach that is much broader in scope than just software reuse or standard compliance. The COE concept encompasses both, but its principles are far more reaching, as it comprises
- from a structural viewpoint
  - an architecture and approach for building interoperable and open systems,

- an approach and methodology for software and data reuse,
- an infrastructure comprised of a set of common standards and strategic products, to facilitate the required degrees of interoperability, portability and commonality in functional-area applications,
- an integration process for achievement of a target operational system,
• from an implementation viewpoint
  - a collection of reusable building blocks, which can "plug and play" to provide services that are common to all applications (e.g. messaging, directory services),
  - a definition of the runtime execution environment,
  - a set of Application Program Interfaces (APIs) for accessing COE components,
  - an environment for sharing data between applications and systems,
• from an integrity assurance viewpoint
  - a reference implementation on which systems can be built, and
  - a toolset for enforcing COE principles and for evaluating compliance against a set of COE requirements.

The COE must be understood as a multifaceted concept. The four major facets (DII-COE, 1999) are
• the COE as a system foundation,
• the COE as an architecture,
• the COE as a reference implementation, and
• the COE as an implementation strategy.

It is apparent that this multifaceted view is complicated to comprehend in its entirety. However, each of those facets is important only for a specific group of personnel. E.g. the system foundation and architecture views are mostly important to IT policy makers and budgetary authorities, whereas a reference implementation and the implementation strategy are focal for the technical project management.

The COE as a System Foundation

The COE is not a system; it is a foundation for building systems and there is only one COE regardless of the target system(s). Building a target system includes combining COE components with function-specific software. System designers select from the COE's set of building blocks (e.g., products) only those which are required for their application. As a result, for different COE compliant systems the basic functions (e.g., communications interfaces, operating system) are either identical (e.g. all systems use an Oracle database) or are selections from a restricted choice of components with well-defined and acceptable interoperability characteristics (e.g. Windows NT or Unix operating system, both providing POSIX compliant services). This implies that target systems utilize as far as possible common components for common functions as the architectural backbone. Additional function-specific software (either developed or purchased) accesses this foundation through well-defined and preferably standardized APIs or interfaces. Implementing proprietary algorithms or substituting components of the foundation by bespoke products would violate COE compliance and potentially jeopardize interoperability, and is therefore prohibited.

COE compliant system design and component selection can be enforced through budgetary mechanisms (i.e. no money for projects, which propose components that violate specific COE preconditions without proper rationale) and through process mechanisms (i.e. before project completion and acceptance, compliance has to be proven by interoperation with a reference system).

The COE as an Architecture

An architecture is defined as the organizational structure of a system or component, their relationships, and the principles and guidelines governing their design and evolution over time (IEEE 610.12). Concerning the COE, one can distinguish three architectural views (JTA, 1999):

- *The Operational Architecture* is a description of the operational elements, assigned tasks, and information flows required for accomplishing or supporting a function. It defines the type of information, the frequency of exchange, and what tasks are supported by these information exchanges. The goal of a COE is to identify 'packages' of operational functions, which can be used as building blocks for a system's specification.
- *The Technical Architecture* defines a minimal set of rules for the characteristics (services, interfaces, standards) of, and the relationships between, the technical COE components. It provides the technical guidelines for implementation of systems upon which engineering specifications are based, common building blocks are built, and bespoke products are developed.
- *The Systems Architecture* is a description of the systems and interconnections providing for or supporting operational functions. It shows how those systems link and interoperate, describes their internal construction, defines the physical characteristics (e.g. connections, location, identification) of nodes and networks, and specifies system and component performance parameters. The implemented systems shall satisfy the Operational Architecture requirements through standards defined in the Technical Architecture.

The relation between those views is depicted in figure 1.

The COE as a Reference Implementation

The COE shall include an implementation of the components defined to be in the COE. A reference implementation is the key to reusability and interoperability. Use of the reference implementation provided as a testbed is required to assure interoperability and is therefore a fundamental requirement for compliance. The reference implementation shall be used as an environment to test the integration and functionality of new products or technologies, while preserving backward compatibility.

The COE as an Implementation Strategy

The COE also is a basis for an evolutionary acquisition and implementation strategy by emphasizing incremental development and fielding to reduce the time to 'market' for the user. This approach also referred to as "build a little - test a little - field a lot", is a process of continually evolving a stable baseline to take advantage

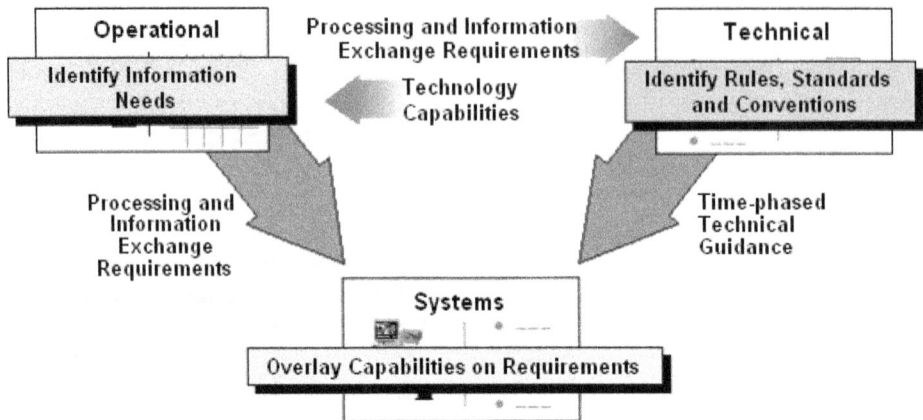

Figure 1 - Architecture relationships (JTA, 1999)

of new technologies as they mature and to introduce new capabilities. Especially in projects, which heavily rely on Commercial-Off-The-Shelf (COTS) products, evolutionary development has become the only practical means, as the traditional development cycle normally exceeds the technical obsolescence cycle. Specification and implementation of large-scale NATO projects with this evolutionary approach, based on some COE core concepts (i.e. mandatory open standards), shows promising results.

The NATO COE Definition

A complete specification and implementation of the above COE requirements and characteristics is obviously an enormous and long-term task. Especially the US have a pioneer role through the already existing, and very successful, implementation of a COE compliant with the above (DII-COE, 1999). NATO started its COE efforts in 1997 and quickly had to face the large complexity of the undertaking. To be able, in the spirit of evolutionary development, to provide useful results early, it was decided to first concentrate on the definition and specification of the standards framework with the major aim to improve interoperability between NATO and national systems. A specification of the runtime execution environment is planned for a later phase. Although the NATO definition already incorporates the runtime execution environment, for the time being the focus is on the standardization perspective.

Definition:

"The NATO COE is the standards-based computing and communications infrastructure, comprised of selected Off-The-Shelf (OTS) products and supporting services, that provides the structural foundation necessary to build interoperable and open systems. The NATO COE will facilitate a common understanding of the concepts, constructs and methods ... required for targeted systems. The NATO COE provides the set of building blocks and guidance

necessary for effective open system design, development, implementation and integration, based, as much as possible, on market proven solutions. It supports an evolutionary systems development approach and provides the executable runtime environment necessary to facilitate the migration and implementation of integrated applications across NATO ... systems."

This chapter will refrain from presenting or discussing specific standards or products, which have been selected within the NATO COE for two reasons. First, the goal of this paper is to present the concept of a COE and not a snapshot on selected items for a specific environment. Whereas the concept of the COE is independent of an organization's structure and context, the components are selected specifically for an organization's tasks and functional requirements. Second, the selection of components and standards is an ongoing process, which would make any detailed market-oriented list of selected items obsolete in short time.

Naturally, a COE in itself can be considered a standard for the organization.

COE BUILDING BLOCKS

From a specification point of view, the COE consists of various building blocks, which provide dedicated information for the developer of a CCIS. These building blocks can provide helpful guidance even outside a COE compliant implementation project, as they are self-standing and independent. When combined through a process, which links them together they can be used directly as a roadmap from the initial operational requirements to the final selection and integration of products.

Operational Requirements

Operational requirements are the starting point for any development of a system. They shall, from a user's point of view, define what functionalities the final system will provide. Two types of requirements can be distinguished:

- Functionality requirements specify the functions needed within a component or system by the user. They describe, on an abstract and implementation independent level, the types of data, the operations performed on these data, the formats for input and output, the performance desired etc.
- Interoperability requirements specify the behavior and characteristics of (sub-)systems in relation to other (sub-)systems. They define the ability of systems to provide and accept services from other systems, and to use those services to enable them to operate effectively together. Interoperability naturally comprises external interfaces for data interchange, but also issues like people-portability and data-reusability.

From a COE perspective, interoperability requirements are an important aspect. In order to structure and classify interoperability requirements, several taxonomies were proposed within NATO for the structuring and automation of exchange and interpretation of data. The currently accepted approach proposes four degrees of interoperability as follows:

- *Degree 1: Unstructured Data Exchange* comprises the exchange of human-interpretable unstructured data such as the free text found in memos, reports and papers.

- *Degree 2: Structured Data Exchange* comprises the exchange of human-interpretable structured data intended for manual and/or automated handling, that requires manual compilation, receipt and/or message dispatch.
- *Degree 3: Seamless Sharing of Data* comprises the automated sharing of data amongst systems based on a common data exchange model.
- *Degree 4: Seamless Sharing of Information* is an extension of degree 3 to the universal interpretation of information through data processing based on cooperating applications.

Obviously these degrees are too coarse to enable or even support standards selection, therefore they have been further refined into functionally oriented sub-degrees, that identify specific interoperability services.

Operational Components

Sub-degrees of interoperability shall not only enable a classification of requirements with finer granularity, but they shall also allow the association of those sub-degrees to operational components (e.g. e-mail) or functional domains (e.g. logistics).

Table 1 shows a subset of the NATO interoperability sub-degrees. Naturally such a list is subject to enhancement and revision based on new or modified user requirements.

Typical or common operational interoperability requirements can be easily described as profiles over the sub-degrees.

As a simple example, a network based system for automatic exchange of formal (EDI-like) messages, with some associated management functionality can be expressed as profile 1A-2AB-3A. This then subsequently would lead to the appropriate standards and associated products as described below.

Standards

A further dimension for the establishment of a COE is a broad overview, analysis and eventually selection of base standards, which provide guidance for a system's composition and product selection. The chosen approach is to select all standards, which promise to be relevant for the achievement of interoperability or associated functionality (e.g. portability). This 'shopping basket' of standards then has to be structured and further assessed and refined, in order to become usable. Several aspects are taken into consideration:

- Structuring by domains or areas of applicability (e.g. networking) to allow quick identification of those standards that might support requirements in a specific domain.
- Structuring by standard priority. To allow for employment of COTS products and to enable broadest possible interoperability, the emphasis should be to prefer standards with widest market acceptance. The proposed categories, in decreasing order of priority, are:
    - Organization specific standards (e.g. NATO STANAGs), although normally not widely supported by the market, have to be considered first,

Interoperability Degree	Sub-degree	Description
1. Unstructured Data Exchange		
	1A. Network Connectivity	Networking services (according to ISO/OSI layers 1 to 4) and file transfer.
2. Structured Data Exchange		
	2A. Enhanced Informal Messaging	Services for informal multimedia electronic mail and associated directory services.
	2B. Network Management	Services for network monitoring and management.
	2C. Graphics and GIS	Services for geo-data, overlay formats, and symbology.
3. Seamless Sharing of Data		
	3A. Formal Messaging	Includes services for formal automated messaging (e.g. EDI).
	3B. Security Management	Multi-lateral security services (e.g. PKI).
4. Seamless Sharing of Information		
	4A. Distributed Applications	Services for distributed computing, object interfaces and object middleware.

Table 1 - Examples of interoperability sub-degrees

as they have been specifically developed and agreed on for an organization-specific purpose,
- International standards (e.g. ISO, ITU),
- Standards from important independent standards bodies (e.g. W3C, The Open Group),
- National standards (e.g. ANSI, DIN),
- Standards defined by industry consortia, and finally,
- Proprietary standards (e.g. of individual vendors).
- Structuring by applicability, maturity and relevance for the desired COE. Proposed categories are:
  - *Mandatory*, where the specified standard has to be used in any case.
  - *Restricted* or *Conditional*, comprising a set of standards, where conditional selection of at least one of those is mandatory.
  - *Emerging* standards, which are foreseen to become relevant in the near future, but that have not gained sufficient market support yet.

This list of base standards shall also serve as a reference book and standards dictionary for project managers and integrators, to provide them an overview, basic explanations and detailed references for potential employment of standards. This document could also be of benefit to organizations that are building non-COE based systems, but seek to improve interoperability by employment of open standards and to lower life-cycle costs.

For NATO the 'NATO Open System Environment (NOSE)' (NOSE, 1998) provides the above standards information. For the purpose of structuring the standards and to classify them by following a reference model, the NOSE has been divided into 11 domains or service areas:

- *Software Engineering Services* provide standards for programming languages, methods and tools (e.g. for CASE) appropriate to the development and maintenance of applications.
- *User Interface Services* define how users may interact with an application. Standards comprise graphical user interfaces, associated look-and-feel and the necessary toolkits.
- *Data Management Services* are concerned with the definition, storage and availability of data, and cover three elements: Data Administration, Repository Control and Database Administration.
- *Data Interchange Services* provide support for the interchange of data (e.g. Graphics, text files, audio/video, GIS data) between applications on the same, or on heterogeneous, platforms.
- *Graphical Services* provide interfaces for manipulating and programming applications concerning images and graphics in a device independent manner.
- *Communication Services* define communications standards, their profiles and protocol stacks structured along the ISO/OSI seven-layer reference model.
- *Operating System Services* provide support for applications, users, devices and other service areas by means of interfaces (APIs). They also describe system behavior and internal communication mechanisms.
- *System Management Services* cover standards for coherent configuration and monitoring of communication systems and networks.
- *Internationalization Services* allow a user to define, select, and change between culturally related application environments, and include character sets and data representation, cultural convention services, and natural language support.
- *Security Services* cover standards for secure communications (on network and operating system level) and associated security management functions.
- *Distributed Computing Services* provide for the extension of local procedure calls to a distributed environment; e.g. for distributed file and print services, distributed transaction processing, or distributed objects.

Standard Profiles

Based on a list of standards, as described above, more detailed profiles are required to establish a system's architecture. Profiles refine each service area into one or more functional classes, with each class mapping to one or more mandatory or emerging standards. Profiles describe those standards in terms of options, parameters, and guidance for product selection and integration.

The Common Standard Profile (CSP) specifies the minimum set of communication and information technology standards within the above listed service areas,

Service Area	Functional Class
Operating System	Kernel Operations
	Non Real-time
	Real-time
System & Network Management	TCP management
	OSI management

Table 2 - Example of service areas and functional classes

to be mandated for the acquisition or major upgrade of all CCIS systems. It focuses on mandating only those standards critical to interoperability, and is based primarily on open system technology, which has strong support in the commercial marketplace.

Some rules apply for the usage of the standards in the CSP:
- If a system has to implement services covered by one of the service areas, it has to adopt the relevant standards mandated in the CSP.
- Specification and usage of other standards than those mandated in the CSP must be additive, complementary, and nonconflicting with CSP mandated standards.
- Legacy standards can be implemented as necessary on a case-by-case basis, on the same premises as in the previous rule.
- New technology can be exploited by the adoption of emerging standards, which may be expected to be elevated to mandatory status when implementations of the standards mature and availability on the market is assured.
- Mandatory standards have priority over emerging standards.

Starting with an initial set of standards, technical discussions and potential consensus on an analysis and presentation of perceived benefits (e.g. of additional standards) by the group which defines the organization's COE, result in further refinement of the CSP. For NATO, the standards in the CSP have been selected on the following criteria:
- *Maturity* of a standard is assumed, when it is technically implemented and when the underlying technology is well understood, robust and tested. A standard is considered not mature when it is implemented with relatively new technology, is not well defined, or when restrictions or problems are known for the standard.
- *Availability* of a standard depends on the level of adoption of the standard by vendors, and on the implementation of the standard within different products. A standard is considered available if two or more products exist that implement the full standard and if those products are available from different vendors.
- *Stability* depends on the advancement or changes expected or planned for a standard. A standard is considered stable if no significant changes are expected or planned within the next two years. A standard is considered less stable if incompatibility exists between the current version and expected or

planned releases.

Mandatory standards, in the sense of the CSP, shall provide the required interoperability and shall also be mature, available and stable.

The CSP is both a forward-looking and living document, i.e. it guides acquisition and development of new and emerging information systems, thus providing a baseline towards which existing systems should evolve. It represents those standards that shall be used now and in the future, and will be updated periodically. It is therefore a document, which will evolve as a result of changes in requirements, technology and the commercial market. As a side effect, NATO's CSP shall also communicate NATO's intent to use open systems products and implementations, and help in influencing the direction of IT industry's standards and standards-based product development.

Strategic Products

The next step in providing building blocks for the COE is to identify strategic products, which fulfil certain operational requirements and at the same time are compliant with the associated mandatory standards of the CSP.

Exclusivity is not required, i.e. it is possible to have multiple different products for the same purpose on the list. Firstly this gives the user or system integrator a choice to utilize products, which he for some reason prefers; and secondly this allows a real-world evaluation and selection of best-of-breed products. For reasons of configuration control and reduction of interoperability risk, the choice should be restricted to a maximum of two or three similar products.

Strategic products support one or more of the three service domains within a COE:

- *Kernel Services* (also referred to as the Minimal Baseline) are that subset of the COE, which is required for all workstations and servers. As a minimum, this subset would consist of the operating system, windowing software, security services, installation software and an executive manager for the runtime environment.
- *Infrastructure Services* directly support the flow of information across various systems. They provide an implementation independent set of integrated capabilities that the applications will access to evoke COE services to move data through the network. Infrastructure services offer capabilities from the following major areas: Management Services, Communications Services, Distribution & Object Management, Data Management Services, and Presentation Services.
- *Common Support Application Services* provide software services that are necessary to view or share data in a common way across the (networked) system. Those services are the glue, which allows and promotes interoperability between various Functional Area Subsystems (FASS). Examples of Common Support Applications include widely used end-user applications such as Office Automation, Online Help, Geographic Services, and Messaging. Common Support Application Services shall be seamlessly integrated through APIs, or at least common file-formats, with the other COE Components.

The selected products shall, in conjunction with their underlying standards,

ensure or at least improve (JSP450, 1998) the following characteristics:
- *Application-portability* is the ability of software to be run on differing hardware and/or operating systems, and to use differing network protocols for external connections.
- *People-portability* is achieved by providing a common look and feel Graphical User Interface (GUI), and by focusing on particular market-leading applications. This enables skills to be transferred from post to post rather than having to retrain on new interfaces and applications to perform, essentially, the same tasks.
- *Scalability* is achieved when a software application is designed to enable it to operate over a range of platform sizes and to run in small, medium and large multi-user environments.
- *Value-For-Money* can be achieved through Economies of Scale and by using market-leading products. Significant cost reductions can be realized through the adoption of a market-led approach to standards; it allows to procure COTS products from many different suppliers, and at different times, with a reasonable degree of confidence that the systems will interoperate and make information available in a consistent and reliable manner. However, it has to be recognized that even established COTS packages are not necessarily free from defects and that adopting a market-led approach can mean commitment to an upgrade cycle that is beyond the control of the organization.

The Linkage of the Building Blocks

The various components of a COE, as described above, imply a need to link or associate these components into a coherent model for the development of a COE compliant system. The necessary decision flow, as depicted in figure 2, is:
- a) Operational requirements of the future system have to be analyzed to determine, which interoperability sub-degrees are providing the necessary functionalities to cover most of the requirements. The result is a functional interoperability sub-degree profile, which identifies the building blocks in a short form (e.g. as described above 1A-2AB-3A). A mapping back to the interoperability degrees is not necessary. The highest sub-degree in the profile (i.e. 3 in the above example) shows the overall interoperability degree of the system.
- b) To identify the standards of the CSP, which support the desired interoperability profile, a mapping has to be established. The basic mechanism is a table showing, which CSP service classes provide support to which sub-degrees. Determining the necessary list of CSP classes and associated standards is a straightforward process.
- c) From the list of strategic products those have to be identified, which provide the required functionality, according to the interoperability sub-degree, and which comply with the mandatory standards determined above. If multiple products are proposed, the integrator has to make a choice based on some project-related rationale. Care has to be taken to select products, which are compatible with the necessary Common Support, Infrastructure, and Kernel Services applications. Typically the composition of a COE has to be per-

CSP Service	Class	1.A	1.B	2.A	2.B	2.C	3.A	3.B	4.A	4.B
User Interface	GUI			X	X	X	X	X	X	X
	Look & Feel				X	X	X	X	X	X
	Toolkit							X		X
Data	Dictionary/Directory				X		X		X	
Management	DBMS (relational)				X				X	X
	DBMS (object-oriented)								X	
	Distributed Data					X			X	X

Table 3 - Example mapping of CSP services and classes to interoperability sub-degrees

formed bottom-up, e.g. by first selecting the Kernel Services, then the Infra-structure Services etc. This shall ensure a strong and harmonic foundation, on which interoperability with Common Support and FASS applications can be achieved easily and with minimal project risk.

d) Eventually it shall not be forgotten, that the above described selection of products has to be validated by some integration (e.g. in a dedicated testbed) and testing to prove the successful coexistence and functioning of those products.

MANAGERIAL ASPECTS

Without proper management efforts, the COE will not be able to efficiently guide the development of large-scale systems, nor will it evolve and become accepted from its initial stages of definition. Major COE management aspects are:

- Setup of a life-cycle model for COE compliant systems, comprising templates for design, implementation, integration/validation, distribution, deploy-

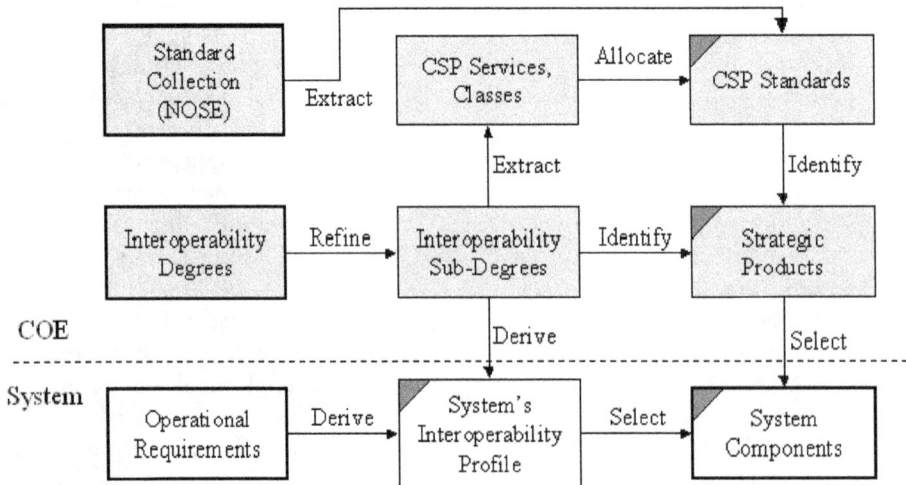

Figure 2 - COE decision flow

ment, upgrade, and, not to forget, disbandment.
- Ongoing review and maintenance of system's compliance with the COE specifications throughout the system's life-cycle:
    - Verification of activities against COE life-cycle model.
    - Incremental implementation in accordance with the COE component model.
    - Compliant integration of new interoperability features (e.g. for interoperability with new external systems).
    - Migration of legacy systems towards compliance.
- Setup of a management structure devoted to the coordination of all COE compliant systems, which is in charge of
    - COE related activities during the system's design, implementation and operational phase,
    - Organization-wide mid- and long-term policy definition (e.g. for acquisition guidance).

Two major aspects of COE management are highlighted and discussed below.

Life-Cycle Management

As for every operational system, the management of the life-cycle of a COE comprises certain phases.
- In the *Initiation Phase* an analysis has to be performed, which shall lead to decisions
    - on how the COE development process will be structured,
    - which major components shall be considered (e.g. only standards, only products, a mixture),
    - what the overall philosophy is (e.g. COTS or development based), and
    - who the participants, partners and sponsors are.

Especially an early involvement of the budget authorities and the future user community (i.e. project managers) is compulsory. One key decision is the allocation of authority and responsibility to COE staff and project leaders. Typically the engineering and testing of COE specifications will be performed centrally, whereas its application, e.g. through product selection, will be done de-centrally by the project staff.
- The *Engineering Phase* comprises the efforts of the partners to generate the COE components (e.g. documentation, testbed, runtime environment, tools, and associated configuration control). Here the major decisions on technical level (especially the mandate of standards and products) are made. To keep up-to-date with technological and market developments, and changed operational requirements, the engineering phase shall be foreseen as a repetitive activity with a predetermined frequency (e.g. forcing a technical COE review once a year).
- The *Distribution Phase* encompasses the effective rollout of the COE to the project and budget authorities. At that point they receive the COE documentation, get access to the testbed and its test results (e.g. on strategic products), and can utilize training facilities and support from the COE staff. As the COE distribution is a natural and tightly coupled follow-on activity of the COE

engineering, it has also to be repeated after every COE review and revision.

- In the *Operational Phase*, the COE as specified is applied to actual projects. Results and experiences of those projects shall be fed back into the engineering phase to improve the overall COE quality.
- The *Disbandment Phase* marks the replacement of the current COE by a major upgrade, or even a different approach to system architectures and project support. The critical aspect is not to destroy the COE infrastructure, as there will be legacy COE systems, which can not be disbanded immediately and therefore will still need the available documentation and tools.

Compliance Testing

COE compliance testing and validation is essential to assure interoperable systems. Testing and validation has to be performed on multiple levels:

- On the CSP level it has to be ensured that the mandatory standards in each service class are nonconflicting and effectively supporting each other, or that the appropriate provisions and restrictions are properly documented.
- Also on the CSP level, all proposed mandatory standards shall fit into reasonable protocol and application stacks, to allow a coherent integration and interoperation throughout the target system's architecture.
- The proposed strategic products shall be tested for compliance with the standards they are supposed to support, to cater for the vendors' varied implementation and interpretation of the agreed standards.
- The strategic products have to be tested for their cooperation with each other, to avoid having products, which formally comply with certain standards, but do not interoperate due to probable noncritical deviations in standard implementation or parameterization.
- And finally, it is necessary to perform regression testing of the relevant parts of the above, when changes to standards or upgrades to products shall be incorporated into the COE.

Typically such a test and validation effort has to be performed on a testbed, where all mandated standards are available in a reference implementation, the strategic products are available and can be installed, and a complete COE compliant system can be rebuilt. Also from the testing viewpoint the desire for a restriction of the COE specification to the bare minimum is obvious, thereby avoiding excessive requirements for testing of interrelations of all potential components.

A further formalization of the validation effort for COE compliance of products and systems would be to follow the approach taken in the DII-COE (1999). Here compliance is rated in eight levels from level 1 (e.g. system uses same standards for operating system, windowing, and database queries), up to level 8 (e.g. no duplication of functions with COE or co-hosted applications; no duplication of data with COE or COE-based target system; 100% compliant with runtime specification, 100% style-guide compliant). Such a detailed view requires a very well defined runtime system, with associated automatic tools for integration and validation, and, implicitly, a long experience and huge effort in building such a demanding environment.

CONCLUSION

A COE provides the functionality and support services necessary to effectively develop, distribute, and integrate interoperable systems across an organization. The COE is scalable, as well as flexible. It embraces a standard-based "Plug-and-Play" open architectural approach that enables functionality to be easily and evolutionary added to a system. The effect of decomposing the system's software into controllable building blocks reduces the time required to put new functionality into the hands of the end-user. And this without sacrificing quality, or incurring unreasonable program risks or expenditures. With the provision of a decision flow from the formulation of operational requirements, over the identification of applicable standards to the selection of strategic products, a most valuable tool is given to the designer of an IT system. Additionally, when strictly obeyed, this approach gives a justification and a proof of value-for-money to the budget authorities, which eventually have to subscribe to the acquisition or development of products. In reverse, an organization's management needs to define a process to enforce the COE compliant system implementation. Typically this would be realized through the approval mechanism for new projects (e.g. technical controlling).

ENDNOTE

1. This paper reflects solely the opinion of the author. It does not necessarily reflect the opinion, policy or intentions of the author's employer.

REFERENCES

ADAGE (1999), Avionics Domain Application Generation Environment. Retrieved from the World Wide Web: http:// www.owego.com/dssa

DII-COE (1999), Defense Information Infrastructure (DII) - Common Operating Environment (COE). Washington, DC: DISA. Retrieved from the World Wide Web: http://spider.osfl.disa.mil/dii

JSP450 (1998), Defence Communication and Information Systems (DCIS) Standards Guides (JSP450). London: MOD UK. Retrieved from the World Wide Web: http://www.mod.uk/dgics/dcsa_web/dseg_web/index.htm

JTA (1999), DOD Joint Technical Architecture (JTA). Washington, DC: DOD. Retrieved from the World Wide Web: http://www-jta.itsi.disa.mil

NOSE (1998), NATO Open Systems Environment. Brussels: NATO. [Retrieved from the World Wide Web: http:// www.nc3a.nato.int/ppdiv/nose/nosehead.htm

Acknowledgments

Special thanks for their support and contributions to El Wells (Mitre), Dr. Fred Moxley (DISA), Wouter Konings (NC3A), Wolfgang Bauer (NACMA), and the members of the NATO COE Working Group.

Chapter XI

The Standardization Problem in Networks - A General Framework

Falk v. Westarp, Tim Weitzel, Peter Buxmann and Wolfgang König
J. W. Goethe-University

INTRODUCTION

Every interaction and all coordination in economic processes is based upon communication. The exchange of information necessarily requires both the sender and receiver of a message to use a mutual language or set of communications standards. Communications standards can be generally defined as rules which provide the basis for interaction between actors (man, as well as machine). These rules must be known or determined ex ante, i.e. before communication begins. If n actors bilaterally agree to a set of communications standards, then n•(n-1)/2 rules must be defined. Such a Babylonian cacophony of languages rarely leads to an efficient exchange of information, however. The uniqueness of communications standards lies in their solely bilateral functionality; they work only when both the sender and the receiver of a message use identical or at least compatible standards. This basic principle for the use of communications standards applies to natural languages as well as to EDI (Electronic Data Interchange) for the electronic transfer of business documents or to network protocols like TCP/IP, for example. Thus, the decision to implement a standard is necessarily tied to the standardization decision of the communications partners. The user's benefit from a given standard generally increases with the number of other users. This phenomenon— the increase of utility derived from a good as the number of users increases—is known as a positive network effect, or demand-side economies of scale, in economic terms. See for example Katz & Shapiro (1985) or Farrell & Saloner (1988). Network effects lead to the interdependence of decisions regarding communications standards by otherwise completely independent actors. This interdependence results in coordination problems because actors, such as firms, do not know in advance when which standards will be implemented by other firms, if at all.

Our research focuses on the examination of alternative forms of coordination and their impact on the selection of communications standards. We have developed two models for evaluating different coordination designs, differentiating between centralized and decentralized coordination of standardization decisions. These models can also be used to analyze and evaluate further cooperation forms between participants in communication networks. We based the economic parameters of the models and the discussion of their implications on empirical research. To gain information about the use of software standards in enterprises, we conducted a comprehensive empirical survey, both in Germany and the U.S.A. It focuses on the corporate adoption and use of various Information Technology (IT) standards, including Internet and electronic commerce standards, business software and EDI.

NETWORK EFFECTS AS THE MOTIVATION FOR STANDARDIZATION

The network effects described above form the basis for an economic analysis of the use of standards. These effects describe the positive correlation between the diffusion of a standard and the benefits for its users. Katz & Shapiro (1985) differentiate between two levels of network effects. Direct network effects describe the direct physical effects which the number of users has upon the utility of a standard. For instance, using a particular EDI standard becomes more valuable the more business partners use the same standard. Indirect network effects arise from interdependencies in the consumption of complementary goods. Thus, widespread use of a standard can be expected to lead to an increased supply of complementary products. The software and consulting services surrounding a new technology like the Internet are one good example. Most studies approach standards as a diffusion of technological innovation, analyzing the circumstances which lead to the spread of different standards. Such work focuses on compatibility. The use of compatible technologies allows any user to become a member of various communications networks, making possible the access to databases and software, the exchange of documents and data, and the simple act of direct communication. Compatibility — and thereby standardization — is then a question of cooperation which has strategic importance for competitiveness (Besen & Farrell, 1994, p. 117). Because compatibility, or standardization, can only be observed in relation to other actors, markets in which compatibility is an important product characteristic are also inevitably markets with strong network effects.

Because network effects are realized by the consumer, the size of the network, and thereby the number of users or the total expected number of participants, is decisive in defining the utility of a standard. No customer wants to buy a technology which does not become the dominant standard, and risk either paying the cost of switching to a different standard or accepting the disadvantages of a small network. Developing and influencing consumer expectations therefore takes on added significance, as expectations rather than the actual product quality or price can determine the decision to buy a specific technology. Like a self-fulfilling prophecy then, poorer technologies can beat out better ones to become the dominant standard, simply because of expectations (e.g. Farrell & Saloner, 1985; Farrell & Saloner, 1986; Katz & Shapiro, 1986; Katz & Shapiro, 1992). Because the producer of a technology which sets the standard can hope to reap abundant profits, manufacturers attempt to gain as large an installed base (number of users) as possible, often through very low prices in the early diffusion phase. From the perspective of the entire network there can be over- as well as under-standardization. A dearth of standards, or excess inertia, can occur when heterogeneous preferences lead to user groups that are too small compared to the collective usage of common standards (Dybvig & Spatt, 1983; Kindleberger, 1983). In contrast, Katz & Shapiro (1986) show than an overabundance, or excess momentum, can occur when producers with market power set very low prices for a standard at early stages, subsidizing early buyers and realizing higher returns from later consumers of the eventually more valuable technology. The existence of network effects can also cause the market to select the wrong number or wrong type of network goods. Because the size of the network or installed base determines the benefits of a standard, it is possible for an existing technology of poorer quality with a large, locked-in installed base to block the success of an innovative, technologically superior standard (e.g. Farrell & Saloner, 1988). Coordination of decisions on the implementation of communications technology among all actors to ensure unified adoption of standards throughout the entire network would therefore seem to be desirable, as every actor thereby realizes the maximum benefits of the network (see Katz & Shapiro, 1986, p. 824). The costs of coordination however, hinder such efforts. The existence of such costs in both the design and diffusion phases of a standard can generate a free-rider problem because there is a positive incentive to avoid the costs of development of and agreement on a given standard by implementing these results free of charge after their evolution. Another negative result of coordination costs is the start-up problem, in which all actors perceive an incentive to wait and see which standard prevails in order to avoid the risk of a premature and possibly unfavorable selection.

THE STANDARDIZATION PROBLEM: A BASIC MODEL

There are various benefits and costs to implementing standards. The common use of IT standards generally simplifies transactions carried out between actors or eases the exchange of information between them. While the use of IT standards can lead to direct savings resulting from decreased information costs due to cheaper and faster communication, standards often induce more strategic benefits and allow the realization of further savings potential. In short, avoiding media discontinuities eliminates errors and costs. The case of 3Com described later will show how significant the potential savings in this particular area can be. The immediate availability of data allows an automation and coordination of different business processes, e. g. enabling just in time production. As a result, an enterprise can reduce its stocks drastically, capital investment in stocks decreases, it can faster react to changes in its competitive environment. In addition, standardization can enhance the exchange of information so that more and better information can be exchanged between communications partners. Because information provides the foundation for any decision, better information implies better decisions. Economically, this is represented as an increase in the information value. As we will later empirically substantiate, the implementation of an IT standard, on the other hand, is accompanied by the costs of hardware, software, switching, and introduction or training — in short, standardization costs. Furthermore, the interdependence between individual decisions to standardize occasioned by network externalities can yield coordination costs of agreeing with market partners on a single standard. More generally, coordination costs embody the costs of developing and implementing a network-wide communications base comprised of a specific constellation of standards which considers the individual, heterogeneous interests of all actors. Concretely, these include costs for time, personnel, data gathering and processing, and control and incentive systems. Depending upon the context, these standardization costs can vary widely (Westarp, Buxmann, Weitzel, & König 1999).

As shown, the fundamental significance of standards lies in their indispensability in the exchange of information. Communication between various actors can be visually described as a network. A communications network is a directed graph without isolated nodes. The nodes represent the communications partners (**i**) (e.g. human, machine, firm), characterized by their ability to process, save and transfer information. The network edges represent the communications relationships. In our models the nodes of a network represent the costs of standardization (K_i) for the respective network actors i while the edges show the costs of their communications relations (c_{ij}) with their respective partners. These costs include the above mentioned costs of information exchange, as well as opportunity costs of suboptimal decisions. Because information provides the foundation for decisions in all areas of the firm, better information implies better decisions. From an economic perspective, this can be seen as an increase in the value of information (see Laux, 1995, p. 289-310). Cost reductions can be realized only when both communicating nodes **i** and **j** have introduced the same or a compatible standard. This does not mean that no costs whatsoever occur to transfer information when both nodes are standardized. Rather, the c_{ij} can be interpreted as the difference between the information costs before (c_{ij}^b) and after (c_{ij}^a) standardization along the respective edge, so that $c_{ij}^b - c_{ij}^a = c_{ij}$. With explicit regard to changes in the information value before (w_{ij}^b) and after (w_{ij}^a) standardization, the information costs savings potential resulting from standardization can be derived from the equation: $c_{ij} = c_{ij}^b - c_{ij}^a + w_{ij}^a - w_{ij}^b$. Thus, the decision problem arises which nodes should be equipped with which IT standard. In our model, there is a tradeoff between the node-related costs of implementing a standard and the edge-related savings of information costs. The benefits of implementing a communications standard must be determined for each node i. In order to do so, the costs of standardization (i.e. node costs K_i) must be compared to the savings c_{ij} to be realized along the edges. If the savings are greater than the costs, then the standard will be implemented. The savings of the edge cost c_{ij} can only then be realized however, if the partner node j also implements this same standard, while the node costs K_i occur independently from the decision of the partner node.

In the case described by figure 1, nodes 1 and 2 have information costs c_{12} and

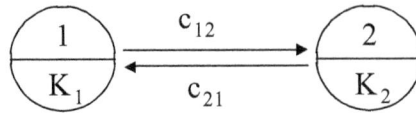

Figure 1. Costs of nodes and edges

c_{21}, respectively. If node 1 or 2 standardizes, it pays the respective standardization costs K_1 or K_2.

If both nodes implement the same standard, they save c_{12} and c_{21}, respectively. If these standards are not compatible however, nodes 1 and 2 pay the costs $K_1 + c_{12}$ and $K_2 + c_{21}$, respectively. In this simplified situation with two actors, coordination of the decision leads to a total benefit of $(c_{12} + c_{21}) - (K_1 + K_2)$ and prevents the firms from paying standardization costs without realizing cost savings (in the amount of c_{12} and c_{21}). The more actors involved and the more different standards available, the more difficult this agreement becomes and the less likely the coincidental, completely uncoordinated implementation of a favorable constellation of standards becomes. Our empirical data show, for example, that very few of the largest enterprises use more than one or two EDI standards.

From the perspective of the entire network standardization is advantageous when total savings on information costs exceed aggregate standardization costs. This approach implicitly applies a collective or centralized utility function as a measure of the quality of decisions. We refer to the standardization problem from the perspective of a central decision making unit (e.g. the state or a parent firm, credited with the aggregate results) as the *centralized standardization problem*. In those cases in which autonomous actors make standardization decisions and are credited individually with the results and responsibility for the effects of these decisions, however, this collective measure at the aggregate level of the entire network is unsuitable. The optimization of the individual objectives of each actor with respect to the implementation of communication standards in the absence of a central, controlling unit is described by the *decentralized standardization problem*. Both approaches describe extreme perspectives in the consideration of coordination mechanisms, providing the basis for examination and evaluation of various hybrid forms of coordination. For a more detailed introduction to the models see Buxmann, Weitzel, & König (1999).

Centralized Coordination

In order to approach the centralized decision problem, we introduce a binary indicative variable x_i, which takes on a value of 1 when node i is standardized and 0 when not (no investment). If i is standardized (that is $x_i = 1$), then standardization costs K_i occur. The standardization costs for the entire network are described by $\sum_{i=1}^{n} K_i x_i$. For information costs, the binary variable y_{ij} takes on a value of 0 if both nodes i and j incident to edge <ij> are standardized ($x_i = 1$ and $x_j = 1$). This leads to the following formulation of the objective function OF^c (c for centralized):

$$OF^c = \overset{n}{\underset{i=1}{Y}} K_i \; x_i + \overset{n}{\underset{i=1}{Y}} \; \overset{n}{\underset{\substack{j=1 \\ j \neq i}}{Y}} c_{ij} \, y_{ij} \qquad \varnothing \qquad \textit{Min!} \qquad (1)$$

$$\text{s.t.:} \qquad x_i + x_j \bullet 2 - M \, y_{ij} \quad \forall \, i, j; \, i \bullet j \qquad (2)$$

$$x_i, x_j, y_{ij} \quad \{0, 1\} \qquad \forall \, i, j; \, i \bullet j \qquad (3)$$

Equation (1) describes the costs of a standardization decision. For $x_i = 1$ *and* $x_j = 1$, restriction (2) in combination with the objective function (1) requires that $y_{ij} = 0$. Our model for solving the centralized standardization problem can calculate the optimal allocation of standards to all actors in any given communications network (first-best solution).

Centralized coordination ensures that a first-best solution is achieved, as all relevant data are considered in the calculation and the means of enforcing implementation of these results exist.

However, in real life application the central manager might encounter significant coordination problems. The coordination costs described above result from the following problems:

- *Data problem*: Given the realistic assumption of asymmetric information distribution and opportunistic behavior on the part of actors, it is questionable that accurate and complete information can be retrieved from all nodes within the network at acceptable costs considering that the nodes' reporting induces resource allocation.

- *Complexity problem*: The centralized standardization problem is a combinational optimization problem. For the basic problem in a network of n actors as described 2^n different constellations have to be compared in terms of aggregate costs; an extended standardization problem considering q different standards even implies comparing 2^{qn} constellations..

- *Implementation problem*: A centralized coordination system is no guarantee that a given, preordained solution will be implemented by all actors. Limits on authority exist even within a hierarchy. The existence of agreements and contracts does not automatically ensure their acceptance. Systems of control and incentives for bridging this problem also involve costs.

Decentralized Coordination

The model described above for solving a centralized standardization problem implicitly assumes that the data, complexity, and implementation problems are resolved (without costs) and that a central manager exists, who can calculate and implement an optimal result network-wide. Given autonomous actors and a realistic knowledge of the data, however, the decentralized standardization problem is foremost a problem of anticipating the standardization decisions of others. Every node i must predict the behavior of all other nodes j (j=1,...,n; i•j). This prediction provides the basis for the decentralized standardization problem. Modeling the decision problem, it is assumed that all nodes i know the various standardization costs occurring in every other node j. This assumption is not particularly restrictive, as these costs can be expected to be similar or at least easily estimated, even for dissimilar firms. It is further assumed that all nodes i know the costs along the edges which directly affect them, i.e. the information costs c_{ij} which they themselves pay and the costs c_{ji} assumed by their direct communications partners. Further data, like the information costs between other nodes, are unknown, as an estimation would be either too inaccurate or too expensive. Assuming that all other nodes standardize, so that savings of c_{ij} are certain, actor i will also standardize when the following statement is true: $\sum\limits_{\substack{j=1 \\ j \ne i}}^{n} c_{ij} - K_i > 0$.

However, node i does not know the strategies pursued by nodes j ex ante, and vice versa. We assume that the actors are risk neutral decision makers. The expected utility of standardization can then be calculated as

$$E\,[U(i)] = \sum\limits_{\substack{j=1 \\ j \ne i}}^{n} p_{ij}\,c_{ij} - K_i \qquad (4)$$

where p_{ij} describes the probability with which actor i believes that node j will standardize. If $E[U(i)] > 0$ then actor i will standardize. If actor i were certain of the behavior of his communications partners, p_{ij} would take on a value of 0 or 1. The decentralized model, however, implies uncertainty. Given our assumptions about the availability of data, p_{ij} can be heuristically computed as follows. Every edge <ij> with costs c_{ij} contributes to the amortization of the standardization costs of the incidental node i. Because the standardization costs K_j and the information costs c_{ji} are the only costs regarding j known to node i, actor i can assume that the edge <ji> is representative for all of j's edges. Combining all assumed data, node i can then develop the following probability estimate p_{ij} for the probability of standardization in node j by attempting to imitate j's decision making behavior:

$$p_{ij} = \frac{c_{ji}\,(n\text{-}1) - K_j}{c_{ji}\,(n\text{-}1)}$$

The numerator describes the net savings possible through the standardization for node j, assuming that all nodes standardize and that the edge <ji> is representative of all of node j's communications relationships (best case). The denominator normalizes the fraction for non-negative K_j as a value from 0 to 1. Should the fraction have a value less than 0, that is $c_{ji}\,(n\text{-}1) < K_j$, then $p_{ij} = 0$ holds. This suggests the following equation:

$$E\,[U(i)] = \sum\limits_{\substack{j=1 \\ j \ne i}}^{n} \frac{c_{ji}\,(n\text{-}1) - K_j}{c_{ji}\,(n\text{-}1)}\;c_{ij} - K_i = \sum\limits_{\substack{j=1 \\ j \ne i}}^{n} p_{ij}\,c_{ij} - K_i \qquad (5)$$

As long as the individual actors are unable to influence the standardization decisions of their communications partners, they can do no more ex ante than estimate the probability that their partners will standardize. *Ex post* of course, the communications costs either remain or are no longer applicable, but the situation of uncertainty described here results from the assumption of limited knowledge of available data. The decentralized model allows the prediction of standardization behavior in a network, thereby creating a basis for predicting the effects of various concepts of coordination. Such measures for influencing the decision to introduce standards also generally apply to influencing the development of expectations regarding the future spread of standards (installed base) or to forms of cooperation which would allow partners to jointly reap the profits of standardization through partial internalization of network effects.

CENTRALIZED VS. DECENTRALIZED COORDINATION

The standardization behavior found in a decentralized network under the above outlined assumptions can be modeled through numerical simulations. Relevant costs were randomly generated for a network of twenty nodes and used to calculate results for function (5). In the simulation, information costs are varied and assumed to be normally distributed with a standard deviation of $\sigma=200$; standardization costs are also normally distributed with $\mu=10,000$ and $\sigma=1,000$. As the results from both forms of coordination are identical when the information costs are either very high or very low relative to the costs of standardization, leading either to complete or absolutely no standardization, respectively, this study focused on more interesting 'moderate' values. To compare the quality of decision making in both extreme forms of coordination, figure 2 shows the results from randomly generated decision making constellations, given a constant expected value of K_i and a continuously increasing expected value of the information costs. Cost savings for the entire network resulting from the decision to standardize are graphed against alternative expected values $E[c_{ij}]$ on the abscissa. In keeping with the above assertions, the graph shows that the results deviate from those of centralized coordination only for more moderate values. These values yield far less in savings. In contrast to the savings realized through standardization under centralized coordination, for values of $E[c_{ij}]$ between 530 and 865, failure to standardize under decentralized coordination results in no change whatsoever in costs. For values between 866 and 1,058, savings for the decentralized network become slightly negative, meaning that total costs increase.

From the perspective of the entire network, that means that the wrong decision was made, because the deteriorations in the individual positions of each node outweigh the improvements. In contrast, wrong decisions are impossible under

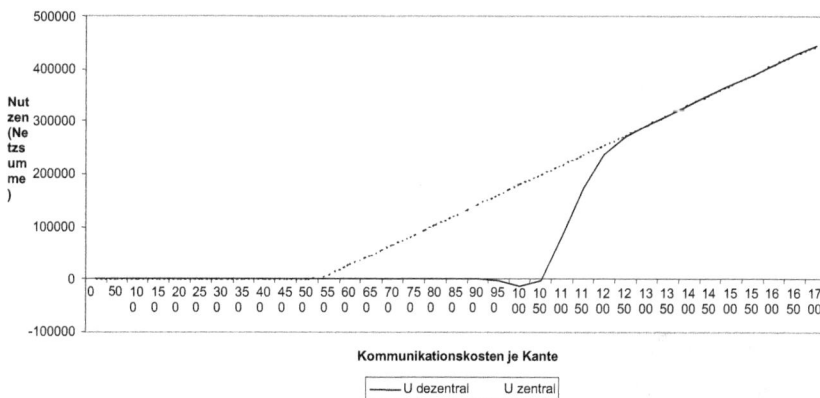

Figure 2. Quality of decisions in a decentralized model based upon ex post network-wide savings on information costs for alternative $E[c_{ij}]$

centralized coordination, as the ex post results cannot deviate from the ex ante planned results. Should the costs of standardization increase, nodes would not have been standardized. Measured against centralized coordination, too few actors standardize under conditions of decentralized coordination. In the range from 1,059 to 1,281, 40.6% to 99.75% of all actors standardize, yielding savings over the entire network which begin to approach those achieved under centralized coordination (in general, with network-wide standardization). The savings are equal where $E[c_{ij}] = 1,282$ with a 100% standardization rate.

Narrowly interpreted, these correlations are only valid for the constellations of parameters upon which they are based. Structurally however, other values and distribution assumptions yield analogous results. Generally, it can be said that given decentralized coordination, the frequency of standardization ceteris paribus increases with c_{ij} or a growing installed base and decreases with K_i. The risk of incorrect decisions under decentralized coordination leads to less willingness to standardize within the moderate range of values. Measured in cumulative, network-wide costs, the accuracy of the solution achieved in the centralized model cannot possibly be attained by the decentralized network. On the other hand, there are no coordination costs in the decentralized model. Using the heuristic described above for computing p_{ij} decentralized coordination yields the same results above and below this middle range as the generally more costly centralized coordination. Network-wide savings attainable through centralized coordination determine the critical value for the costs of coordination above which a centralized solution is no longer advantageous. This value corresponds to the perpendicular distance between the curves in figure 2.

The models also provide a good starting point for further analysis towards the question of how incentive systems or local cooperation within networks will improve the quality of decisions. These improvements manifest themselves in the realization of greater potential savings caused by "correct" decisions to standardize or the reduction or prevention of "wrong" decisions on standardization. For example, every actor could calculate the individual maximum possible benefits resulting from standardization as the difference between the costs of standardization and the costs along all incidental edges. The actor could use a portion of the positive difference over his current (ex ante) expected value to pay premiums to other nodes to ensure that they implement a standard. In the same way, large firms or firms with high edge-related costs c_{ij} could give away options to their partners which hedge the loss in the case of an incorrect decision. Risk averse actors may also be prepared to pay a hefty sum for such options. The existence of intermediaries in networks offers yet another alternative in the form of outsourcing. An actor who implements multiple standards can also offer such services, thereby reducing its own costs. For a comprehensive model see Westarp, Weber, Buxmann, & König (1997).

EMPIRICAL ANALYSIS

Examining questions of managing IT standards in enterprises we conducted a survey in the summer of 1998. A questionnaire containing about 30 questions on 8 pages was sent to 1,000 of the largest companies both in Germany and the United

States (see for an online-version of the questionnaire http://caladan.wiwi.uni-frankfurt.de/IWI/projectb3/eng/survey). Prior to mailing the questionnaire, each company was contacted by telephone to identify the head of the MIS department to whom the questionnaire was then directly addressed. 250 completed questionnaires were returned in Germany (25%), and 102 in the US (10.2%). The goal of this study was to gain empirical data about corporate adoption and use of various IT standards. On the one hand, the study was designed to provide an insight into the determinants of strategic standardization issues like the diversity of software solutions, compatibility problems, and the centralization of decision structure. On the other hand, more detailed questions, e.g. about benefits and costs, were asked for the selected categories *Internet* and *electronic commerce standards, business software* and *EDI*. To evaluate our standardization models, we will first empirically analyze costs and benefits of standardization decisions in business networks using EDI as an example. In a second step, we will evaluate the standardization problem in enterprises and examine the question of centralization vs. decentralization of decisions.

Standardization in Business Networks: The EDI Example

To enable efficient communication within business networks, the participants often face the decision problem to agree on certain EDI standards. The results of our empirical study show that about 52% of the responding enterprises in Germany and about 75% in the US use EDI technology to transfer structured business data. In average, German enterprises use EDI with 21% of their business partners, while it is 30% in the US.

Enterprises that want to use EDI have to choose from a variety of different standards to structure the content of the documents. The respondents were asked which particular standards are in use in their companies. Figure 3 and figure 4 illustrate what percentage of the responding companies uses the respective EDI standards. For reasons of simplification, we only show a static snapshot of the use of different EDI standards in 1998 (see for the diffusion process over the past 14 years Westarp, Weitzel, Buxmann & König 1999).

EDIFACT by far is the most popular EDI standard in Germany. It is used by nearly 40% of the responding enterprises. Other common standards follow with a large distance. About 11% of the responding German companies use VDA, 6% use

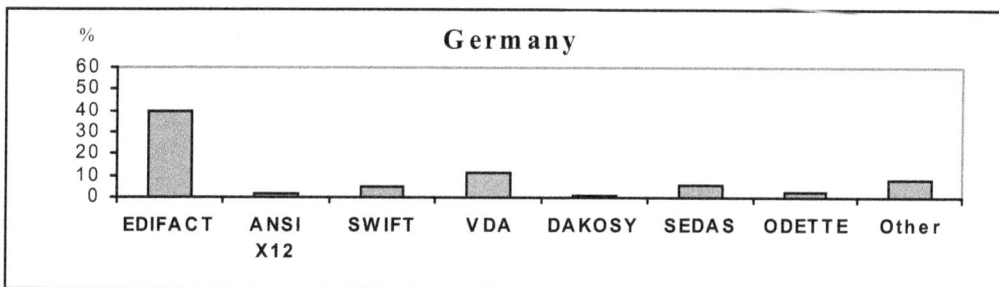

Figure 3. The use of EDI standards in Germany

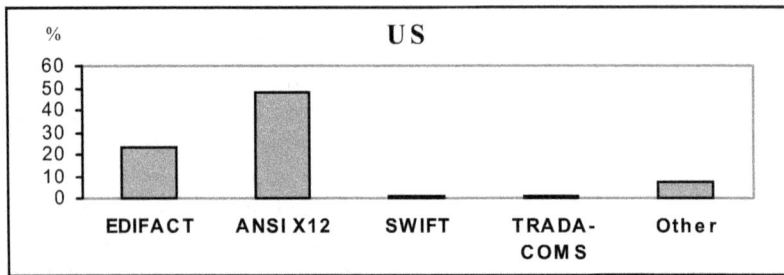

Figure 4. The use of EDI standards in the US

SEDAS, 5% use SWIFT, 3% use ODETTE, only 2% use ANSI X12, and about 1% use DAKOSY. Despite the strong position of EDIFACT, the survey reveals a rather strong heterogeneity in the use of EDI with many different industry-specific standards. The coordination problem is obvious.

This heterogeneity of EDI standards cannot be found in the United States. The leading US EDI standard is ANSI X12 (more than 48%) followed by EDIFACT (24%). Only two of the responding enterprises in the US use SWIFT or TRADACOMS respectively. A reason for this is that the US government influences the coordination of EDI standards (similar to a central coordinating unit in the centralized model) by preferring the ANSI X12 standard for government-related transactions.

To evaluate the parameters of our models we also asked the participants using EDI about the costs and savings of implementing and using an EDI solution. Analyzing the data we found that there was great heterogeneity in terms of realized costs and savings since the context of using EDI varies extremely between different companies (see Westarp, Buxmann, Weitzel & König 1999). We will therefore illustrate empirical figures of costs and savings by studying the case of 3Com (see for a more detailed version of the case study Westarp, Weitzel, Buxmann, & König 1999).

With over $6 billion in annual revenues, 3Com is one of the largest players in the electronics and network communications industry worldwide. Mainly driven by the request of strong US business partners 3Com implemented an EDI solution based on the ANSI X12 standard in 1995. Due to the ongoing globalization EDIFACT was additionally implemented in 1998 since a large number of the international business partners uses this standard. EDI decisions in general are made by the EDI Operations unit, which is part of the highly centralized IS Organization department. Within the different business units there are also EDI analysts who decentrally coordinate the interdependencies between EDI data flow and business processes. 3Com uses EDI with about 15% of its approximately 200 business partners (suppliers and distributors). Redesigning the distribution channel management the company has recently focused on larger distributors that are more likely to support EDI. This explains the fact that already about 50 per cent of the company's suppliers are EDI capable. With EDI becoming more and more a significant strategic issue for enterprises, especially in the computer industry,

3Com tries to convince more of its business partners to use EDI; for new suppliers and distributors it is a requirement already. Focusing on the costs of implementation four areas can be separated.

- The *start up costs* for the EDI solution were less than $25,000 (including the first year of VAN service) since already existing technical and human resources were used.
- With the growing data traffic *new EDI project implementation* took place. Establishing the EDI Operations department, new personnel was hired. Also, the technical infrastructure was upgraded, installing a new Unix translator for $100,000.
- The *setup of a new trading partner* for EDI at 3Com takes about 2-3 days of a programmer's work.
- A *new transaction setup* like adding a certain document to the existing set of an EDI partnership takes a programmer about 8 working days (which is rather low compared to an industry average of about 2 weeks), and the mapping takes about $1,140.

The annual costs of running the solution are estimated at $350,000 for personnel, $36,000 for the data transmission (VAN services), and about $17,000 for additional external services, such as software license agreements and outside contracting consultants. 3Com also uses a significant portion of the budget for continuing education and professional conferences.

Compared to the benefits, the costs of the EDI solution seem to be rather reasonable. At 3Com the costs of manually processing an order process are calculated at $38 compared to $1.35 using EDI. This sums up to estimated savings of $750,000 in sales order and invoice processing. Taking also the reduction of data entry errors, efficiency increases due to better warehouse management, and reduction of processing delays into account, the EDI Operations department estimates overall savings of $1.3 million. These figures are expected to increase dramatically next year since 3Com is in the middle of consolidations due to the merger with U.S. Robotics. At the moment the EDI systems of the two companies are in the process of integration.

Heterogeneity, Compatibility Problems and the Question of Centralization

After empirically examining the costs and savings parameters of our models, we will now evaluate the standardization problem within enterprises by analyzing how heterogeneous software solutions currently are and to what extent compatibility problems arise. We will also examine the question whether centralization of decisions might help to solve such problems.

To gain information about the heterogeneity of software solutions in companies the respondents were asked about the number of different products they have currently in use in the different categories. Figure 5 illustrates which percentage of the responding companies uses how many products in each of the listed software categories. The upper and lower bars of each category show the results in Germany and in the US respectively.

Comparing the two countries, the results do not reveal a significant difference concerning the variety of products. "Business software" is an exception to this.

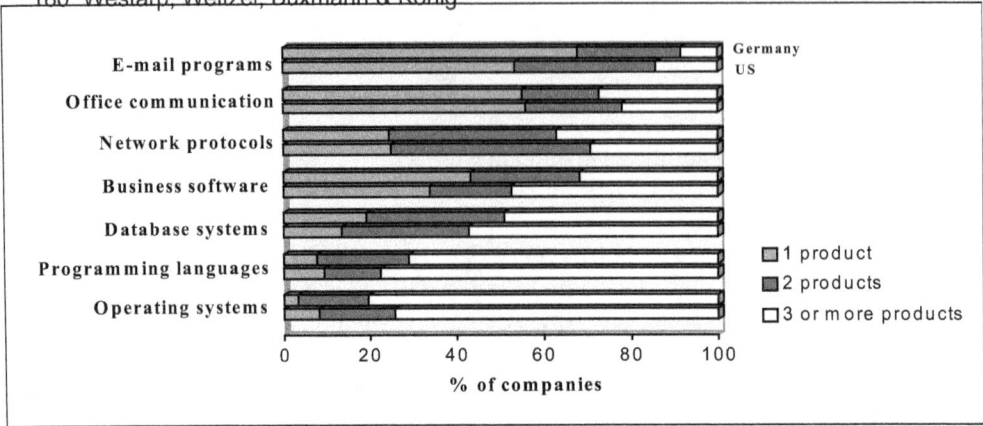

Figure 5. Heterogeneity of software products used in the various software categories.

More than 68% of German responding companies only use one or two products in this category, while it is only 53% of the companies in the US. This remarkable gap of 15% is most probably a result of the dominant market position of SAP software in Germany. The largest variety of standards is found in the area of "Programming languages" and "Operating systems". Only 29% of the largest companies in Germany use less then three programming languages (23% in the US) and only 20% (26% in the US) use less than three operating systems. In contrast, we find less variety in the categories "E-mail programs" and "Office communication". About 92% of the responding German companies (86% in US) use only one or two different products for their e-mail communication and 73% (78% in US) use only one or two different office communication products.

Looking at the variety of standards used does not necessarily provide enough information about related problems. Therefore, the respondents were asked to give their evaluation concerning the incompatibility problems within the software categories classified above. A five category Likert scale was used for this question with the extremes "very significant" and "very insignificant". Figure 6 illustrates how often the categories "very significant" and "significant" were chosen. For reasons of simplification, the figure does not show the answers in the other categories.

In general, enterprises in the United States seem to be more likely to experience problems of incompatibility than in Germany. The largest differences appear in the categories of "Business software," "E-mail programs" and "Programming languages."

MIS managers of both countries report their largest problems of incompatibility in the area of "Business software". This is likely to be a result of the strategic importance of such systems in enterprises. Nowadays, business software has an impact on all key processes. Therefore, incompatibilities in this area lead to more significant problems than in other areas. In general, we find that incompatibilities are an important matter in today's enterprises and therefore, that the standardiza-

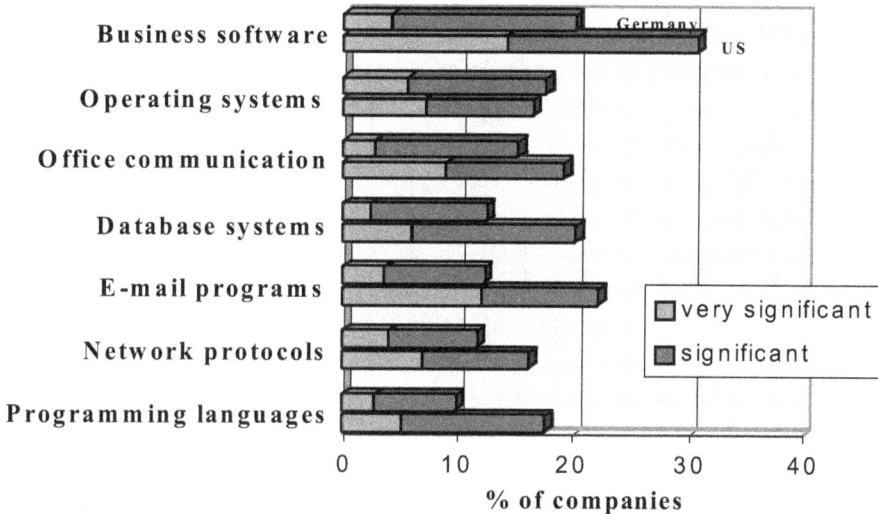

Figure 6. The problem of incompatibility.

tion problem we described in our models is of practical relevance.

We assume that the chance of incompatibility problems increases with the number of different products in use. To test this hypothesis we measured the correlation between these two variables. We have to consider that the variable that measures the number of different standards is an interval variable while the incompatibility problems were measured ordinally. In this case, it is appropriate to apply statistical tests for the ordinal level. Therefore, the interval variable was transferred to an ordinal scale. To do this, the six categories "1 product in use", "2 products in use", "3 products in use", "4 products in use", "5 products in use" and "more than 5 products in use" were created. We then calculated the Goodman's and Kruskal's Gamma and the Spearman's Rank-Order Correlation Coefficient (rho). Both of them are applicable to measure the correlation of ordinal variables (Levin & Fox, 1997). In our case the Goodman's and Kruskal's Gamma is a particularly appropriate measure of correlation since the variables involved are ranked in categories (Levin & Fox, 1997, p. 331). In table 1 we display the measured values of gamma and rho whenever the level of significance (α) is better than 0.05, since a α larger then 0.05 would be statistically questionable.

In most of the cases we see a moderate positive correlation between the two variables. For the German data sample in every software category all levels of significance for both coefficients are better than 0.0005 (SPSS shows the value .000). This means that we can be sure with a confidence of more than 99.95 percent that the measured correlation is not a result of sampling error.

The empty spaces in the table indicate the areas in which α is larger than 0.05 and therefore correlations are not significant. This applies for the categories "Network protocols," "Programming languages" and "Operation systems" in the US sample. One reason for this might be that this sample is smaller than the German one. However, in general we find a positive correlation between the problem of

	Goodman's and Kruskal's Gamma				Spearman's rho			
	Germany		US		Germany		US	
	gamma	α	gamma	α	rho	α	rho	α
Office communication	.301	.000	.371	.003	.244	.000	.305	.004
Database software	.295	.000	.252	.013	.276	.000	.241	.025
Business software	**.459**	**.000**	**.428**	**.000**	**.408**	**.000**	**.422**	**.000**
E-mail programs	.382	.000	.463	.000	.261	.000	.361	.000
Network protocols	.369	.000			.312	.000		
Programming languages	.310	.000			.289	.000		
Operation Systems	.315	.000			.294	.000		

Table 1. The correlation between incompatibility and the number of different standards/ products used.

incompatibility and the number of different software standards used.

There is a particularly strong correlation in the category of "Business software". Taking into consideration that the MIS managers also reported this category as the one with the largest problem of incompatibility, it seems reasonable to reduce potential incompatibilities by reducing the number of different products used in the company. We will use the example of business software to further analyze if centralization can reduce the number of different products in use and therefore reduce incompatibility problems. We propose the hypothesis that centralization leads to more homogeneous software solutions. In order to test this hypothesis for the field of business software, we measured the potential correlations between centralization of decision making and number of different products used for the data samples of both countries.

The respondents were asked to answer the question who makes the decision in the field of business software. To answer this question the respondents could choose from the following categories:

1. company management
2. head/employee of the central MIS department
3. head/employee of an operating department (e.g. controlling, accounting)
4. head/employee of a business unit
5. other

Multiple answers were possible. For analysis and interpretation the data was regrouped. Answers of category 1 and 2 were classified as "central", 3 and 4 as "decentral". Whenever answers were found in both of these groups, they were newly classified as "team". Answers were also counted by the variable "team", whenever it was explicitly mentioned in the category "others" that decisions were made with central and decentral units participating. The table below shows the proportion (in percent) of how decisions concerning business software are currently made.

In most of the cases we find that decisions concerning the selection of business software are made within teams, i.e. both central IT departments and decentralized

in % of companies	decentral	team	central
Germany	3.3	56.7	40
US	8.5	61.7	29.8

Table 2. Who decides about the implementation of business software?

departments are involved in the decision process. An interesting result, however, is that in Germany decisionmaking is currently more often centrally organized.

It appears to be promising to take a closer look at a potential connection between the larger degree of centralization (see table 2) and the smaller heterogeneity (see figure 5) in Germany in comparison to the United States. In accordance with the results of our numerical simulation centralization seems to lead to more homogenous solutions. This correlation appears to be reasonable since a central decision making unit is able to consider company-wide network effects related to the use of software standards.

Based on our findings, we will try to validate this hypothesis empirically using business software as an example. In order to do this, we measured the potential correlation between *centralization of decision making* and *number of different products used* (the latter as an indicator for the degree of standardization) for the data samples of both countries. To measure potential correlation between the two variables we calculated the *Goodman's and Kruskal's Gamma* and the *Spearman's Rank-Order Correlation Coefficient (rho)*. Table 3 shows the results.

While for the German sample no significant correlation was found, testing the

Goodman's and Kruskal's Gamma				Spearman's rho			
Germany		US		Germany		US	
gamma	α	gamma	α	rho	α	rho	α
-.053	.637	-.388	.006	-.036	.633	-.307	.010

Table 3. The correlation between the degree of centralization and the number of different products used.

US sample shows a moderate, statistically significant, and negative correlation between the number of different products in use and the degree of centralization. Therefore, we accept the research hypothesis for the US sample: The number of different business software used in enterprises decreases when the degree of centralization increases.

This result seems to substantiate the findings of the simulation in section 4 that centralization leads to more standardization and therefore to less problems of incompatibility. Nevertheless, the measured correlation needs further examination since it could statistically only be proven in the comparatively small US sample.

CONCLUSION AND FURTHER RESEARCH

Centralized coordination of standardization decisions best describes the situation within a firm or a combine. We introduced a model for solving the centralized standardization problem, which can calculate the optimal assignment of standards to actors in any given communications network. The decentralized model we introduce describes the standardization behavior of independent actors given limited knowledge of relevant data and a lack of hierarchical authority. Numerical simulations show that, in general, a higher degree of standardization is expected in centrally coordinated networks. The failure to realize collective savings potential follows in lockstep. These analytical findings are backed by our empirical data.

Using EDI as an example, we gave an overview of the current structure of costs and benefits of implementing IT-standards in large companies. We showed that the data needed for the evaluation of different coordination mechanisms by our models can be collected empirically. We also evaluated the significance of the standardization problem in large enterprises. We gained data about the status quo of IT standards used (number of different products in use in different categories) and proved statistically that with an increasing number of products the problem of incompatibility is also increasing. This indicates the lack of compatibility and appropriate interfaces among current software products, which leads to the standardization problem we analyze in our framework. For the US sample we were able to prove the positive correlation between centralization and standardization that was already shown by the simulation results.

Using our framework as a starting point, we will conduct further research towards coordination costs and incentive mechanisms in the context of standardization. Furthermore, we will implement the models in cooperation with a large German enterprise as a controlling instrument for standardization decisions. Additionally, we will use relational diffusion models to examine the decentralized coordination of consumers in software markets.

REFERENCES

Besen, S. M., & Farrell, J. (1994). Choosing How to Compete: Strategies and Tactics in Standardization. *Journal of Economic Perspectives*, Vol. 8, No. 2, 117-131.

Buxmann, P., Weitzel, T., & König, W. (1999). Auswirkung alternativer Koordinationsmechanismen auf die Auswahl von Kommunikationsstandards, in *Zeitschrift für Betriebswirtschaft (ZfB), Ergänzungsheft 2(99)*, 133-151.

Dybvig, P. H., & Spatt, C. H. (1983). Adaption externalities as public goods. *Journal of Public Economics*, Vol. 20, 231-247.

Farrell, J., & Saloner, G. (1985). Standardization, Compatibility, and Innovation. *Rand Journal of Economics* 16, 70-83.

Farrell, J., & Saloner, G. (1986). Installed Base and Compatibility: Innovation, Product Preannouncements, and Predation. *The American Economic Review*, Vol. 76, No. 5, 940-955.

Farrell, J., & Saloner, G. (1988). Coordination through committees and markets. *Rand Journal of Economics*, Vol. 19, No. 2, 235-252.

Katz, M. L., & Shapiro, C. (1985). Network externalities, competition, and compatibility. *The American Economic Review*, Vol. 75, No. 3, 424-440.

Katz, M. L., & Shapiro, C. (1986). Technology adoption in the presence of network externalities. *Journal of Political Economy*, Vol. 94, No. 4, 822-841.

Katz, M. L., & Shapiro, C. (1992). Product Introduction with Network Externalities. *Journal of Industrial Economics*, 40, 55-83.

Kindleberger, C. P. (1983). Standards as Public, Collective and Private Goods. *KYKLOS*, Vol. 36, 377- 396.

Laux, H. (1995). Entscheidungstheorie 1, *Grundlagen*, (3rd ed.), Berlin.

Levin, J./Fox, J. (1997): *Elementary Statistics in Social Research*, 7th edition, New York.

Westarp, F. v., Buxmann, P., Weitzel, T., & König, W. (1999). The Management of Software Standards in Enterprises - Results of an Empirical Study in Germany and the US. SFB 403 Working Paper AB-99-7.

Westarp, F. v., Weber, S., Buxmann, P., & König, W. (1997). Communication Services Supplied by Intermediaries in Information Networks: The EDI Example. SFB 403 Research Report FB-97-2.

Westarp, F. v., Weitzel, T., Buxmann, P., & König, W. (1999). The Status Quo and the Future of EDI—Results on an Empirical Study. Proceedings of the 7th European Conference on Information Systems (ECIS'99), 719-731.

Acknowledgments

This work is part of the research project "Economics of Standards in Information Networks" of the interdisciplinary research program "Competitive Advantage by Networking - the Development of the Frankfurt and Rhine-Main Region". We thankfully acknowledge the financial support from the German National Science Foundation. We are very grateful to Tom Trunda, 3Com Global EDI Project Manager, whom we conducted the 3Com case study with. We are also very grateful to Arie Segev and Judith Gebauer of the Fisher Center for Management & Information Technology at the University of California, Berkeley who supported our empirical survey in the US. We very much appreciate the work of Frank Ladner who implemented the web-based version of the questionnaire. Finally, we are very grateful to Martin Frech, Alexander Conzelmann, Astrid Ruhrmann, and Ralph Seibert for the data entry.

Author's Note: For further information on the research project "Economics of Standards in Information Networks" please visit http://www.vernetzung.de/b3/ or contact the authors at the Institute of Information Systems (Institut für Wirtschaftsinformatik), J. W. Goethe-University, Mertonstr. 17, 60054 Frankfurt am Main, Germany, Telephone: + 49 69 798-23318, Fax: + 49 69 798-28585, http://www.wiwi.uni-frankfurt.de/IWI/, {westarp|tweitzel|pbuxmann|koenig} @wiwi.uni-frankfurt.de.

Chapter XII

Standardizing Retail Payment Instruments

Simon L. Lelieveldt[1]
De Nederlandsche Bank[2]

The business of payments and the provision of payment instruments have a rich history, which can be drawn upon in a discussion of standardization. In the middle-ages, for example, the mere existence of a wide variety of foreign and local coins led to a flourishing business of money exchange offices and cashiers in the Netherlands. Malpractices of some of these firms, mostly in the form of physical tampering with coins and alloy, resulted in government regulation on a municipal and province level. Yet, as these type of regulations where hard to enforce, the Amsterdam municipal government decided in 1609 to establish a municipal exchange bank, 'de Amsterdamse Wisselbank', originally as a government monopolist. The motivation for doing so was to prevent the regular price-increases of the good coins, to eliminate confusion to the public and to facilitate trade by providing good coins. Later on, in 1621, the regulations were adapted to the actual business practice and private cashiers were allowed – under certain conditions – to conduct business in the city of Amsterdam (van den Berge, 1939, p 34).

The example shows us how a diversity of specifications and a diversity of payment instruments, will lead to the development of separate companies which make money by reducing the confusion for their consumers. It illustrates that the abuse of technological know-how and abilities for the sake of increased economic benefits by a few private companies may lead to government intervention for the sake of public interest. Furthermore it indicates that strong market powers may prevail, even in the case of restricted government regulation. As such the example contains all relevant issues with respect to IT-standardization:
- can it be assumed that the market will standardize if necessary?
- what role should governments play in this process?
- does the end-user play a role in this process?

In this chapter, I will examine the above standardization issues with respect to the retail payment instruments, developed and in use since the beginning of this century. In this time frame bank notes and coin have been widely available to the public as a basic (and standardized) payment instrument. I will however not include these instruments in this study and limit myself to a study of the standard-ization of noncash payment instruments that have been available to the consumer. These payment instruments can be seen as the technical means with which consumers effect money transfers to each other. Examples of payment instruments are the forms for credit transfers or in-payments, the debit- or credit cards or home-banking software. It is my opinion that, given the availability of cash an alternative payment instrument, the standardization processes of noncash payment instru-ments can be seen as the 'pure' result of market forces. The study of this process, applied to different types of instruments within one application and industry domain, will hopefully provide additional insight.

An important reason for limiting the study of standardization to the develop-ments in the Netherlands is that the retail payment sector is typically a geographi-cally segmented market, with large differences between countries. Even between countries that are physically very close to each other, there are considerable differences in culture, payment habits, legislation and institutional arrangements with respect to supervision of financial markets and regulation of competition (Revell, 1983, p 80). As these differences have a considerable influence on the dynamics of competition within the financial sector, they also have an important effect on the dynamics of the standardization process. Consequently, even a comparison of Dutch standardization developments with those in Germany or Belgium (that have a similar giro-based payment infrastructure) would have to involve a study of the specifics of these different markets and their different institutional settings. I have chosen not to do so and to limit myself to a longitudinal study of the standardization of retail payment instruments in the Netherlands.

The approach taken in this chapter, is an inductive, bottom-up approach. I will try to highlight the elements of the standardization process, which from a practitioner's view are of utmost importance for enterprises, governments and consumers respectively. In doing so I will outline my assumptions and implicit theoretic notions first, and then provide documented examples. I hope that this approach may stimulate the theory development by the scientific community on this subject.

THE BUSINESS PERSPECTIVE

For a good understanding of standardization, it is useful to establish the different nature of:
- payment systems,
- technical standards,
- business standards,
- business agreements.

In the above list, only the payment system can be seen as a fully specified setup of people, machines, procedures and payment instruments which results in the transformation of input (payment instruction) to output (payment). The other items in the list must all be seen as conditions or guidelines, which may somehow effect the operational payment system or guide the design of future payment systems.

Payment systems, which are part of the business process of an organization, are by definition firm specific. Given the characteristics of the organization (centralized or decentralized, using certain processing equipment, assuming certain employee qualifications etc.) the actual payment systems will differ per type of organization. As a result, payment instruments which serve the same purpose for bank A and bank B will have, in many cases, different specifications, geared to the optimal use in those firm-specific business processes. These differences may relate to the account-number structure, the size of the instrument, the option to truncate information etcetera. The effect of these differences is that even small changes in the specification of the actual instrument (use of extra check-digit for example) may involve large changes (and high conversion cost) in the way that the payment instrument is implemented in the business process of the particular organization.

As a result of basic economics, it may be assumed that every provider of payment instruments has standardized its internal processes. Using a similar lay out for all types of payment instructions (such as the credit-transfer, the standing order and the in-payment) of the same institution will for example reduce the number of errors made by customers, using both instruments. It can further allow optimization and standardization of back-office procedures and processes. These in-house technical standards can be seen as guidelines, used to enhance or enforce similar specifications of payment instruments and payment processes within one single institution.

In addition to the in-house standards, competing providers may establish that a certain standard will be adopted as a common guideline for their payment systems. The agreed standard (which can be technical but may also be of a legal or procedural nature) then becomes a business standard. Generally, business standards are established in the context of domestic bankers' association. Examples of those standards, which may cover more than just the technical aspects of payment instruments and processes, are:

- a clearing house arrangement, in which payments are netted multilaterally before being settled between participants,
- a code of conduct with business principles for issuing payment instruments and processing payments,
- requirements as to the minimum content of contract terms,
- the use of a similar layout for certain classes of paper-based payment instruments,
- the use of a similar user interface for certain classes of electronic payments,
- requirements with respect to key management in point of sale systems.

The benefits of agreeing to business standard can be diverse. A cooperative agreement on basic security requirements and certification for bank terminals for example, will allow IT-hardware suppliers to be able to manufacture more standardized equipment for the market, instead of developing separate individual solutions for different banks. Consequently, the market for this equipment may be larger and more open, which will result in lower prices for the equipment. Cooperative agreements on minimum contract terms and redress procedures could serve the purpose of generating trust with the public, by providing clarity with respect to conditions and customer complaint procedures.

In practice only the largest providers of payment instruments have say in the negotiations on the precise content of the business standards. If one of those large providers does not cooperate, a too significant part of the benefits of standardization cannot be achieved. So although the standardization process may formally involve many players, including government agencies, the large providers are actually the most important players. During the process, the large providers share technical information on the variety of solutions that may be allowed or facilitated in the standard. In general the outcome of the process will not be the standard which technically provides the best solution to the problem which has to be solved (which may be interoperability, but could also be the conversion to a different key management structure), but a technical standard which results in solving the problem under the constraint that the implementation of the standard in the firm-specific business processes of the banks amounts to the same magnitude of conversion cost for each of the large providers.

It is important to note that it is almost impossible to arrive at a proper technical standard if the institutions involved have no operational experience, nor a legacy problem, in the application domain of the standard. Only those institutions that have wrestled with the problem of converting a functional description into a working payment system are really able to appreciate the trade-offs that have to be made in the implementation of the standard. Such trade-offs typically result in a preference for a pragmatic, sometimes illogical, technical solution to a logical but inexecutable solution.

The importance of distinguishing a technical standard from the payment system itself is that it helps to understand that an agreement on collectively implementing the same technical standard (with a specific purpose) does not imply that the specifications of the payment systems are the same. If banks agree to use a similar message protocol in their Automated Teller Machines, this does not imply that they would all have to buy and use identical hard- and software and establish exactly the same internal procedures. Each institution will determine their optimal way to comply with the standard, given its current technical and organizational infrastructure. Distinguishing standards from payment systems further helps to understand that it is possible that two payment systems, which are based on the same standard, do not have to be interoperable. Every standard requires some further decisions with respect to the specific use of the standard and the way to fulfil the requirements of the standard through a specific implementation of hardware, software and procedures. Those implementation decisions will be geared towards the infrastructures of the entities involved and will be unique for each issuing bank. Imposing a standard is therefore no magic bullet for achieving interoperability.

Having discussed the business standard, I will now focus on a specific business agreement, which is closely related: the business agreement on reciprocal use of payment instruments (often called: agreement on interoperability). Although the existence of business standards will often go hand in hand with decisions to process other institution's payment instruments, this is not necessarily the case. Further cooperation than standardization is a separate business decision, which constitutes of determining:

- the degree of cooperation; this may vary from acting as a remote mail box (sending all instruments and payments immediately to the issuing bank) to

acting fully on behalf of the issuing bank itself (actually performing part of the processing of the payment),
• the application and level of interbank fees applied to the cooperative activities (based on number of instruments / transactions involved).

Both content and conditions of the cooperative agreement will reflect the underlying market position and bargaining power of the participants to the agreement. Consequently the stronger entity in the bargain (larger market share, greater negotiation skills) may exercise more influence over the outcome.

In general it can be observed that institutions with a similar client base and a similar business process (same rate of centralization/decentralization, same hard/ software platform, same cost/benefit sensitivity) will be the first institutions to agree on standardization and later also interoperability. Similar agreements between institutions that have both commercially and technically a different setup are more unlikely and will involve longer and more complicated negotiations on both standards and fees when bargaining power is equally divided. Smaller providers will have limited market power and may be unable to significantly influence the standardization process or to negotiate sharp fees. Yet, their benefits in entering these agreements originate from the fact that the cost of provision of the related services by the small institutions themselves may be higher than by buying it from larger competitors.

An important factor, determining bargaining power for institutions is the de facto standard, which could be seen as a working system or product which covers a substantial part of the market. In this respect, three important trends can be identified. The first is that the very important role of the IT-supplier has become smaller vis-à-vis the role of the institution that applies the information technology products. The second trend is that the battle for standards is increasingly taking place on an international scale, as some payment products have developed into regional or international products. The third is the increasing role of government institutions, notably the European regulators (Egyedi, 1996). The above trends increase the playing field and the number of players and make it more difficult to understand the dynamics of the standardization process. However, as far as the role of the provider is concerned, the dynamics remain unchanged and can be summarized as achieving optimal efficiency while limiting the necessary change to the business processes.

Case examples

In the years 1900-1925, the use of noncash payment instruments increased considerably. The payment process was largely a paper-based administrative process. Private organizations involved in executing payments were the cashiers and banks (Westerman, 1920). The Municipal Giro Institute of Amsterdam was set up in 1916 and a nationwide Postal cheque and giro-service was established in 1918. In this time-period both Vissering (1907) and Van Vliet (1923) promote the use of a uniform form lay out for executing payment transactions between banks and cashiers in order to execute these transactions more easily and with less errors. Van Vliet (1923) also points out that the use of machines would in the future further increase speed of processing.

From 1925 to 1945, the centralized giro services gained a lot of market share in the retail payment market. Concentration in the banking sector led to a stronger position for a smaller number of nationally operating banks, while regional banks and cashiers slowly lost their business. Wolf (1983) describes how in 1937 the private bankers' association finally adopted a proposal to introduce a uniform payment form, to be used by consumers of the banks. It was agreed that the payment form would not be promoted too actively, in order to prevent cashiers from losing too much business. As for the use of checks, Hammerstein (1998) describes how the Geneva Convention of March 19, 1931 was the basis for the Dutch law on uniform checks, which was applicable since 1934.

After the Second World War, the Postal Giro was the largest player in the retail payments market. When in the 1950s, computer technologies allowed for the introduction of new processing methods and payment order forms, discussion took place between the private bankers and the Postal Giro, to determine if common ground could be found. The differences between the two parties were however too big. Whereas processing of the Postal Giro was centralized and uniform, the processing of payments by private bankers was still too diverse to be able to agree on a single payment form. The Postal Giro therefore decided in 1962 to start using punch cards as the payment form (Wolf, 1983, p 33).

The use of the punch card in a centralized processing structure allowed the Postal Giro to expand its services in terms of volume and customer segments. As a result, the Postal Giro Services accumulated considerable deposits at the cost of the private bankers (who were actually in need of deposits, given the credit-restrictions set by the central bank). The banker's reacted by setting up a banker's giro, developing a unified bankers payment order form (not a punch card), introducing a unified bank number and introducing the guaranteed check for payment at the point of sale system (Wolf, 1983). Also, in order to prevent credit cards from entering the European market, the banks developed the Eurocheque, a guaranteed uniform payment check for cross-border use in Europe (Muns, 1974).

Wolf (1983) describes how for a specific payment product, the inpayment ('acceptgiro'), the private banker's in 1966 decided to adopt the instrument of the postal giro and start discussions on adapting it to the banker's needs later. These discussions led to a preliminary arrangement from 1974 to 1978 (adaptation of the postal giro cards for use by banker's) and a final arrangement (introduction and use of a commonly standardized instrument) from 1978 onwards.

Peekel and Veluwenkamp (1984) describe the characteristics of the centralized processing in the girocircuit of the Postal Giro Services as well as the partly centralized processing in the banker's giro circuit. The standardization that took place to overcome the lengthy time-periods necessary for processing payments between these two circuits concentrated on determining standardized interbank message interfaces in order to allow the institutions to continue to process their payments in their own characteristic way. These interfaces were subsequently implemented for all kinds of payment instructions under the auspices of the steering group National Payments Circuit. The actual design and implementation of these agreements lasted more than twenty years (1975-1998) due to the fact that not only technical but also political and commercial difficulties had to be overcome.

The development of new payment instruments or delivery channels such as automated teller machines and electronic funds transfer at the point of sale shows that differences in processing-structure as well as in cost/benefit structure determined the strategic choices of banks and Postal Giro (since 1986: Postbank). Postbank for example focused at first on substituting check-use for Electronic Funds Transfers at the Point of Sale (EFTPOS) transactions and later on Automated Teller Machines (ATM's). This was due to the branch-contract with Post Offices, as a result of which installation of ATM's would not lead to any cost reduction but to cost increase. Meanwhile other banks invested heavily in the installation and deployment of ATM's. This resulted in a situation in which the Postbank, with the largest market share in number of customers, had the smaller market share in number of ATM's. An interoperability agreement on ATM-use would thus result in large usage of bank-ATM's by Postbank clients and a correspondingly large interbank fee to be paid by Postbank to other banks. Also, in the early stages of development, the on-line to issuer concept of Postbank's ATM transactions was not easily to be implemented in bank ATM's. The actual interoperability agreement on ATM-use was effected more than ten years later in 1998.

The situation for the EFTPOS-application was somewhat different. All large players had a similar market share in terms of processing guaranteed checks by retailers. Therefore, a collective investment in a common product would result in a similar change of cost/benefits for all players involved. Still the banks had to be stimulated by the large oil companies (Shell) and retailers (Albert Heijn) as well as the central bank in order to reach a final agreement on setting up a joint business operation for EFTPOS in 1989. The technical basis for the system was the system developed by the Postbank, which evolved into the interbank de facto standard for EFTPOS.

Similar observations can be made for the development of the electronic purse (Chipknip and Chipper), home banking on the Internet (I-pay) as well as the development of cross-border EFTPOS and ATM-use (edc, Maestro, cirrus). These would show that the international dimension of standardization becomes more important. As a result, interoperability technical business standards for future chip-card based debit- and credit card use have been established (the Europay Mastercard Visa-standard). Also industry initiatives have been setup to ensure that future standards for electronic purses will take into account the local initiatives developed thus far. The most recent initiative in this respect is the plan to develop a standard that defines a PC-chipcard reader for use with financial applications. One of the driving forces behind this latter initiative is to limit the possibility that hard- or software suppliers with a large market share start the provision of less-secure chipcard readers, which may become a de facto standard.

THE GOVERNMENT PERSPECTIVE

First of all, it should be noted that the government perspective does not by definition have to be uniform. Governmental institutions which influence standardization can range from local government authorities, Ministry Departments, supranational institutions, central banks, secret services to national and international standards developing organizations. Each of these institutions has its own motivation for steering and guiding standardization. In addition to these institutions, parliament might also exercise influence on the process by approving or disapproving legislation in this domain. The one thing however that most of these regulators have in common is a lack of implementation knowledge with respect to most domains of standardization. In addition some regulators also tend to make analytical and managerial mistakes, which will be discussed below.

One of the most common mistakes made by regulators is what I would call: solving the problem of the observer. If an observer of payment instruments would list the wide variety of solutions and contract terms available in the market, this may appear to be confusing and may call for standardization. Yet, the existence of this variety of solutions may in itself not bother any user or provider. A user will choose the payment instruments that suit him and will not use all payment instruments available in the market. Similarly the providers may make some additional money by providing integrated or standardized solutions if the variety is too big. Still, some regulators find it hard too deal with the wide variety of solutions in the market and therefore urge for standardization. The actual message they are thus sending out is that solving their observers' problem is more important than understanding the market.

Another mistake made by regulators is to formulate interventions in terms of prescribing technical procedures or measures taken, instead of prescribing the desired outcome. A regulator could for example prescribe that certain maximum processing times for payments would have to be met, without specifying the means to arrive at that goal. Such an approach would leave the market parties to decide whether or not to standardize.

As the effect of regulatory interventions may be that market parties have to bear conversion or implementation cost, regulators should be very careful in the nature of their intervention. Their action might result in an arbitrary, possibly very unfair spread of cost amongst market parties in the same market. If not motivated properly, it can be viewed as a random tax burden, imposed by the regulator. Regulators therefore have to decide whether they operate on the assumption that they have a full understanding of the market and its specific problems and characteristics or on the assumption that they are ignorant and that all entities in the market will try to influence and use them to achieve their individual interests. From a fairness point of view the latter assumption is more adequate than the former. Yet in practice the former appears to be more widely held than the latter.

In essence, the challenge for regulators is to refrain from all interventions, which are different from stimulating market parties to standardize. Any other action should be motivated by a distinct problem in terms of a non-functioning market. Even then, the necessary intervention should not be to impose standards but to prescribe functional requirements that solve the problem at hand, leaving open the option to standardize. Regulators that act differently basically spend the tax payer's money on either their own goals and problems (the observer problem) or on providing a competitive advantage to a market party that best succeeds in influencing the regulator. Although these latter expenditures are beyond the mandate of most regulators, some still find motivations to act this way.

Case examples

The prime example of the observer's problem may be found in the domain of European harmonization. Based on the perception that all different technical systems in the countries of the European Community constituted a problem for consumers and enterprises, the Commission focused on harmonizing technical standards among countries. Although this may have helped to create a level technical playing field, it is still an open question to me if this technical issue is the most important barrier to create an internal market without trade barriers.

An example of the requirement formulation problem can currently be observed in the discussions on electronic commerce and the payment instruments applied. Politicians as well as regulators tend to believe that a technical standard is necessary to ensure the availability of a very safe payment method. This can be viewed as a too technical formulation of the requirement or the desire that market parties start offering payment methods which are acceptable to users (leaving open if the acceptability is reached by security measures, fee structure or contract terms).

A historical example of government ignorance is the attempt of secret services to prevent the Data Encryption Standard from becoming an ISO standard (Lelieveldt, 1989). One of the motivations was that this would prevent the widespread use of DES. Although the attempt succeeded in the end, it failed to have the desired effect as applications of DES were already being developed and applied on the basis of the available FIPS standards.

Another example of the same problem is the sponsoring by the European Commission of projects and standardization of a common European electronic purse. As a matter of politics (one electronic euro-purse as well as one physical euro) the goal is understandable. From a business perspective however, there is no business case. Only some 5 % of retail payments are cross-border ATM and EFTPOS transactions, which provide the consumer with sufficient cash or payment options. As a result, within Europay the decision was initially made to not standardize the electronic purse. The initiative to develop a Common European Purse Standard (CEPS) was taken up however in reaction to the announcement of Visa, which stated to develop a worldwide purse. Given its political goals, the European Commission became a willing sponsor for this activity.

Examples of process interventions can be observed in the developments of the National Payments Circuit and the EFTPOS system in the Netherlands. Both Ministry of Finance and the central bank stimulated market parties to reach agreement. Similarly the central bank declared to favor a deployment of electronic purses which would make technical differences transparent to the user. As for the interoperability of ATM's no regulatory pressure has been exercised on banks and Postbank to reach an agreement.

THE USER PERSPECTIVE

Although in economic theory the user perspective plays an important role in determining the success of products and services, it should be noted that payments do not constitute a primary need, but are essentially a derived need. Both Aders (1984) and van der Have (1972) describe banking and making payments as a convenience or an experience good as opposed to a search good. Once the choice has been made for one or the other bank, the use of this bank becomes a habit. Its services are used often and the customer has little motivation to spend time and effort in reconsidering this choice. A strong motivation is needed to change the initial choice. This motivation can be for example a serious problem with the financial services rendered or a 'role change'. Van der Have describes these role changes as getting the first job, getting married, moving to another house, owning a car, moving in with the partner etcetera. These occasions may give rise to changing the former habit.

In practice, the choice of a bank account is still very much determined by the proximity of the bank branch (which ensures easy access) and by the choice already made by the parents of the client (which ensures familiarity with and trust in the services offered). Consequently, the strategy of banks towards retail consumers is focused on increasing the number of services offered to current clients, which increases the burden of changing to another bank. As for future clients of the bank, the focus is to approach those at switch moments in their lives (moving out, getting a job etc.) starting with approaching the very young consumers.

Aders (1984) points out that the homogenous character of bank services and the importance of proximity of the branch to the consumer leads to a situation of homogenous oligopoly. In combination with the fact that retail payment and banking services are a convenience good, the effect of price of specific services on consumer behavior is rather limited. The role of standardization as a factor of consumer choice is even smaller. The primary choice factors -within the range of instruments available in a certain purchasing or payment situation- are risk, control of payment moment and convenience. Of course standardization helps to increase the convenience for consumers, yet convenience and ease-of-use can also be achieved through other means (self-explanatory form layout and interfaces).

As stated earlier, the efficiency considerations of providers will lead to a level of in-house and inter-organizational standardization that is good enough for users. As standardization is not a significant consumer choice factor, providers do not have to consider other than internal efficiency considerations in determining the necessary degree of standardization. Consumers won't switch from bank A to bank B because an instrument is not standardized. In cases where the diversity would really become a burden, some providers will definitely start providing integrated or standardized solutions.

Case examples
In practice, most forms and methods used within a bank are standardized. Also, for the most important external interfaces such as the direct debits, inpayments, standing orders, dialogues for electronic funds transfers and cash withdrawals, business standards are agreed and applied. Yet, the payment form used by the Postbank differs substantially from the forms used by other banks. Even though some 60 % of the population have an account at both Postbank and another bank, standardization has not taken place. Similarly, home-banking applications, Internet-applications and voice-response applications are not standardized between banks. From a technological perspective, two different types of electronic purses are available on the market (Chipper and Chipknip) and where this might pose a problem for the merchant, this is solved by providing a combined terminal which

accepts both types of cards, not by standardizing the chipcard-purse itself. On an infrastructure level, the Dutch banks investigate, however, the development of a standard for a chipcard reader, to be used with the Personal Computer for all sorts of chipcards, including those for financial applications and transactions over the Internet.

CONCLUSION AND RECOMMENDATIONS

Perhaps the most important lesson that can be learned is that the characteristics of a specific market have a large impact on the dynamics of the standardization process. As payments are a derived convenience good, the importance of standardization to the consumer is rather limited. The dynamics of standardization in this area are therefore heavily influenced by the dynamics of competition between providers in this sector. In this process, providers try to achieve optimal efficiency by adopting technical standards and agreeing on business standards, while limiting the necessary change to the business processes. Important trends that can be observed are the diminishing role for the IT-provider as the supplier of the de facto standard, as well as an internationalization of the battle for standards between providers. A final trend is the increasing role of government institutions, notably the European regulators.

In my opinion the factor which most complicates the standardization process, as far as the market of payments is concerned, is that many regulators tend to assume that standardization can be seen as a useful regulatory tool. Those regulators fail to recognize that they lack the implementation experience in the relevant domain, as well as a good understanding of the dynamics of competition in the market. As such, regulators may become tools in the hand of influential market parties instead of the tools of the taxpayers. Furthermore the priority of regulators may be biased towards solving observer problems and perceived political problems instead of trying to contemplate whether the market at hand is sufficiently competitive (in which case standards will evolve).

So if we are to look at the IT-standardization process and how it can be optimized in theory, the most important recommendations concern the behavior of companies and regulators. In general regulators should refrain from using standardization as a regulatory tool, as it can be assumed that in a well functioning market standardization will take place as a matter of economics or non-standardized solutions will be available to solve any diversity problems in the market. Companies might want to try to be more open in explaining their motivation for joining and influencing certain standardization processes. This will provide regulators as well as users with a better view of the market and the intention of its players.

In practice however, companies and regulators will have few external incentives to change their current behavior as the consumers and citizens lack the knowledge as well as the market and political power to really affect the behavior of those institutions. Any change in the current standardization practices of companies and regulators therefore rests on the intrinsic motivation of these organizations to shape their responsibilities to the consumer or the citizen in an appropriate way.

ENDNOTES

1. The author would like to thank E. Michiels (Moret) and J. Gigengack (ING Group) for their comments and suggestions.
2. This chapter reflects the professional opinion of the author and may not be interpreted as a policy position of De Nederlandsche Bank.

REFERENCES

Aders, J. H. J. (1984). *Marketing van betaaldiensten.* Amsterdam: NIBE

Advokaat, H. G., Have, J. van der & Pauwels, F.L. (1972). *Retailbanking in Nederland.* Amsterdam: NIBE.

Berge, L. G. van (1939). *Giroverkeer in Nederland.* Den Haag: Uleman.

Egyedi, T. (1996). *Shaping standardization.* Delft: Delft University Press.

Hammerstein, E. (1998). *Betalingsverkeer (wissel, orderbriefje en cheque).* Deventer: Kluwer.

Lelieveldt. S (1989). *Elektronisch betalen goed geregeld.* Rotterdam: Lelieveldt.

Muns, E. C. (1974). Banken in vier landen beginnen met het eurochequeproject. *Bank- en Effectenbedrijf* 23 (164), 125-128.

Peekel, M., & Veluwenkamp, J.W. (1984). *Het girale betalingsverkeer in Nederland.* Amsterdam: postgiro/rijkspostspaarbank.

Vissering, G. (1907). *Het oude en het moderne giroverkeer.* Amsterdam: J.H. De Bussy.

Revell, J.R.S. (1983). *Banking and Electronic Funds Transfers.* Paris: OECD.

Vliet, J. H. van (1923). Normalisatie van het incasso- en girobedrijf bij de bankinstellingen. *De bedrijfseconoom* 2 (5), 99-103.

Westerman, W. M. (1920). *De concentratie in het bankwezen.* 's-Gravenhage: Martinus Nijhoff.

Wolf, H. (1983). *Betalen via de Bank.* Amsterdam: NIBE.

Chapter XIII

Institutional Constraints in the Initial Deployment of Cellular Telephone Service on Three Continents

Joel West
University of California, Irvine

INTRODUCTION

The influence of institutional pressures on standards and standardization are readily apparent in their most direct form. For example, in the mid-1990s, both the European Union and the United States issued new wireless communications licenses in the 1.8-2.0 GHz band: the EU countries mandated use of their decade-old communications standard, while the U.S. authorized three competing standards not yet widely used in the U.S. (Mehrotra, 1994).

However, institutional pressures can also shape standardization efforts in a less direct fashion. For example, in a regulated industry such as telecommunications, existing economic and political institutions constrain the diffusion of a new technology. Such diffusion mediates the impact of product compatibility standards upon society. If producers adopt standards for their goods and services, and if users adopt the products that incorporate such standards, only then such standards can have an economic or social effect upon society at large. Therefore, it is important to understand the impact of institutional pressures on diffusion of the innovation that incorporates a standard if we wish to explain the eventual success or failure of such a standard.

Here a particular standards-based innovation, analog cellular telephone service, provides an opportunity to contrast the effects of institutions on diffusion and thus standardization. Over a four year period, three independent design centers deployed mutually incompatible standards in three continents. While the technical solutions were similar, differences in institutional context between the regions influenced both the nature of the respective standards and their corresponding diffusion. In particular, the systems were deployed in a period of shifting telecommunications competition policies and priorities for radio frequency allocation.

Prior research has examined the causal links between standards and institutions, both the institutional context of standards development (e.g., Besen, 1990) and also how established standards themselves function as institutions (Kindleberger, 1983). But rarely do we have the opportunity to examine the diffusion of the same innovation in differing institutional contexts.

This paper will focus on the most complex institutional context for the deployment of cellular telephone service, the United States, which despite having invented cellular technology, was the third region to deploy cellular service due to regulatory delays. The experience of Japan and Northern Europe are offered as contrasts to highlight the importance of the institutional context in the adoption of both standards and standardized products.

DIFFUSION OF STANDARDIZED TECHNOLOGY PRODUCTS

Institutional Context of Standardization

Although sometimes viewed as merely technical in nature, product compatibility standards are tightly interwoven with economic and political institutions, as they can be both the consequence and antecedent of such institutions.

Standards normally originate in institutions, whether economic, political or a hybrid thereof. Economic standards-setting institutions can be either a single firm or a coalition of firms, while political institutions sponsoring standards may be national, regional and international governments. Other standards originate from hybrid organizations, committees of individuals or firms to whom the government delegates responsibility, such as the American National Standards Institute (Farrell & Saloner, 1988; David & Shurmer, 1996).

Many standards stem from institutions whose scope is the nation-state. Some standards — such as those for broadcasting — are implicitly dependent on governmental institutions for the arbitration of competing claims and the promulgation of uniformity within a national market. Strong national institutions may also be a prerequisite for the adoption or substitution of standards, such as the standardization of railroad gauges. Global standards may originate with such national standards, or developed by explicitly multinational institutions such as the International Telecommunications Union (Kindleberger, 1983; Besen & Farrell, 1991).

Whatever their source, standards themselves serve as economic institutions: they fit the class definition of a public good, available to all and not depleted through use. Indeed, developing such public goods often falls to government by default (Kindleberger, 1983; Cowan et al., 1991; Antonelli, 1994). These economic institutions can constrain industry structure, defining the basis for both vertical supply relationships (e.g., Intel to IBM) and horizontal competition (IBM vs. Compaq).

These two aspects of institutionalization — institutions driving standards and standards acting as institutions — are often joined in the *ad hoc* institutions that are inevitable for multi-vendor *de facto* standards. Such institutions seek to promote the common goals of standards adherents while dampening their inevitable competitive rivalries (Gabel, 1987).

"Technology Push" Diffusion Strategies

The pattern for deployment of cellular telephones highlights an ongoing debate over the appropriate balance between two alternatives for diffusing technological innovation: developing solutions in response to perceived demand, or offering new solutions based on what has become technologically possible. The "technology push" approach used for launching cellular technology was consistent both with other discontinuous innovations, and also with the institutional constraints under which telecommunication companies operated during this period.

While contrary to accepted normative rules of product marketing, the deployment and diffusion of radically new technology-based products and services have for decades have been based on a "technology push" strategy (Levitt, 1960). Mowery and Rosenberg (1979) long ago observed the timing and availability of new technologies are more often determined by supply availability rather than demand pull. Dosi (1984: 9-10) notes the inadequacy of "market pull" strategies in explaining radical innovation: "the range of 'potential needs' is nearly infinite and it is difficult to argue that these would-be demands can explain why, in a definite point in time, an invention/innovation occurs."

Key problems in doing market research for radical innovation come in the assumptions of the eventual target market. As Lynn et al. (1996: 11) observe, "the familiar admonition to be customer-driven is of little value when it is not at all clear who the customer is." Even after the first customers, the early feedback can be misleading given the major differences between initial enthusiasts and the early majority (Moore, 1995). This failure to correctly identify the ultimate customer was present in all of the initial cellular markets, although some operators and manufacturers were quicker than others to recognize the eventual mass market.

The tendency towards technology push strategies can be accelerated not only by the nature of the innovation, but also the institutional framework under which it is deployed. In the century since the invention of the telephone, monopoly carriers in the developed world commonly deployed new telecommunications services using a technology push strategy. In most countries, these monopoly carriers were government-owned Post, Telephone and Telegraph (PTT) companies, organized as a government department or state-owned enterprise; in a few countries (such as the U.S. and Canada), the carrier was a regulated private firm.

New telecommunications offerings thus faced a path markedly different than for unregulated (and competitive) technological industries such as personal computers or software (Langlois, 1992). For telephone companies, the diffusion of innovations was institutionally constrained either through the PTT's government sponsors, or through government regulation of private telephone companies. The deployment of cellular telephone service was a clear example of such a "technology push" approach, particularly given that demand was consistently underestimated in the U.S. and Japan (Noda, 1996: 178; Funk, 1996).

At the same time, crucial differences in the regulatory frameworks between regions contributed to early differences in the diffusion patterns. The initial deployment of cellular telephone service in the U.S. (when contrasted to Europe and Japan) starkly highlights the impact of regulatory institutions upon supply-push technological innovation. The necessity for exclusive allocation of a scarce resource — radio frequency spectrum — provides the government with absolute control over market entry by cellular telephone service providers.

DEPLOYMENT OF CELLULAR TELEPHONE SERVICE

Evidence of institutional moderators and path dependencies in standards adoption can be seen in the development and deployment of cellular telephone service in the United States, particularly when contrasted to parallel developments in Japan and Europe.

The cellular telephone concept was invented by AT&T's Bell Laboratories in the 1940's, but the U.S.' first fully-licensed system did not become operational until 1983. Three major factors accounted for this delay: maturation of the technology (supply), the changing role of AT&T and the long-delayed FCC decisions on allocating frequency spectrum. As we will see, the latter two factors help account for why the U.S. deployment lagged other nations — despite its role in inventing the technology.

Technology Supply: Development of Mobile Cellular Telephony

AT&T launched the world's first commercial mobile telephone service in 1946. In developed countries, such systems were operated in large cities for a decade or more prior to the introduction of cellular telephones (Young, 1979). These initial car telephone systems used one set of frequencies for an entire metropolitan region (thus limiting capacity) and also required manual operator connections (increasing the cost per call). What we now recognize as cellular mobile telephony reflects the combination of two subsequent technological advances:

- *switched mobile radio,* which allowed customers to directly dial outgoing connections to the public switch telephone network: in the U.S., this began in 1964.
- *cellular radio,* which subdivides an urban area into smaller geographic cells, allowing the same frequency to be reused within an urban area. Although a Bell Laboratories scientist invented the cellular concept in 1947, AT&T did not put it into service until 1983 (Young, 1979; Seybold & Samples, 1986).

Despite early recognition of the key concepts necessary for cellular telephony, the technology necessary to implement commercial cellular telephone systems did not become available until the 1970s. Three key developments were:

- *Frequency synthesizers.* Early mobile telephones used oscillator crystals, and were limited to a dozen pairs of transmit /receive frequencies. To increase capacity, additional frequencies were need to avoid interference between cells and to increase trunking efficiency within cells. These problems were solved beginning in the early 1970s, low-cost frequency synthesizers allowed telephones to support a large number of frequencies, e.g., the 832 channels used by the AT&T's Advanced Mobile Phone Service (AMPS) (Young, 1979;

Calhoun, 1988; Macario, 1997).

- *Digital Switching.* Computerized telephone exchange switches (not available until the mid-1970s) were essential for the complex requirements of mobile telephone service, including billing, finding the mobile terminal for incoming calls, keeping calls as the mobile moved between cells, and allowing customers to use their telephones ("roam") outside their home service territory (Meurling & Jeans, 1994).
- *Microprocessors.* Nearly as much computing power was required in the mobile telephone as in the land-based switch. The solution came with the 1971 introduction of the microprocessor. The car telephone of AT&T's 1978 Chicago field trials used the same Intel 8080 microprocessor used in the Altair 8800 — the personal computer credited with launching the U.S. PC revolution (Fisher, 1979; Langlois, 1992).

By the mid-1970s, the key technical obstacles were overcome: these technologies were sufficiently mature to support large-scale urban cellular systems. All were incorporated by telecommunications operators in the U.S., Japan and Northern Europe as they developed their respective cellular technologies — but U.S. deployment lagged due to two institutional factors: the changing role of AT&T and the FCC consideration of frequency spectrum for mobile telephony.

Liberalization and AT&T's Declining Role

In the U.S., cellular telephone technology was developed by AT&T and, to a lesser degree, Motorola. As such, the deployment of cellular telephone service in the U.S. must be understood in the context of AT&T's traditional role, which evolved during the telecommunications liberalization that culminated in its 1984 breakup.

The Bell Company was organized in 1877 by Alexander Graham Bell and others to license the Bell's patents, and then integrated horizontally and vertically during the next 50 years. The resulting "Bell System" (1925-1983) linked three national resources — Bell Labs (research), Western Electric (manufacturing), and AT&T Long Lines (inter-exchange telephone service) — to some two dozen local operating companies, which were monopoly utilities regulated by individual states. AT&T accounted for the majority of U.S. local telephone service and nearly all long distance revenues (Chandler, 1977; Temin, 1987; Friedlander, 1995).

AT&T was accused by rival telephone manufacturers of using its service monopoly to monopolize the equipment market. In 1949, the Justice Department filed an antitrust lawsuit to force divestiture of Western Electric and require the Bell System to buy equipment by competitive bidding. In settling the lawsuit with the 1956 Consent Decree, AT&T retained its vertically integrated Western Electric subsidiary only by restricting its manufacturing and operations to common carrier telephone services (Temin, 1987). In particular, it forfeited the right to build and operate private mobile radio systems, thus limiting its experience with FM radio propagation — a crucial core competence needed in the subsequent development of fixed and mobile cellular telephone equipment (Young, 1979; Calhoun, 1988).

Facing no direct competitors in a regime of gradually increasing demand and declining costs, AT&T and its subsidiaries had virtually no layoffs and below-average turnover. This stability facilitated a strong corporate culture that was driven by service rather than marketing. Unlike most companies of comparable size, powerful marketing executives were unheard of until the 1980s. For local operating companies, the leading executives before World War II were engineers that had built the system, while postwar executives were primarily from customer relations ("commercial") and operations ("traffic") divisions (Feldman, 1986; Temin, 1987).

As with monopoly PTT's in other countries, the incentives for the local operating companies neither encouraged nor rewarded risk-taking and innovation. Feldman (1986) concluded that the guaranteed rate of return system gave managers no incentive to increase revenues and profits. Protected from competition but not from the losses of overexpansion, risk-averse monopoly managers delayed deploying unproved new technology until demand developed. So the technology push did not come from the local companies, but from Bell Laboratories. Arguably the U.S. leader in corporate basic research from 1945-1983, Bell Labs invented the transistor, laser, communications satellites, among other technologies. It continued to develop the cellular concept, demonstrating an experimental system to FCC regulators in 1962.

But the role of AT&T — then the world's largest corporation — changed dramatically in the 25 years leading up to 1984. Temin (1987: 7) argues that "the process was dominated by changing ideology, not changing technology," an ideology that later spilled over into FCC decisions about mobile telephony. It was also reflected in the decisions (1959-1977) by federal courts hearing challenges to AT&T's dual monopolies on equipment provisioning and inter-exchange service. The courts granted firms like Motorola, Carter Electronics and MCI the right to directly compete in AT&T's previously monopoly markets (Kahaner, 1986; Temin, 1987).

The last phase of change began in 1974, when the U.S. Justice Department filed an antitrust suit alleging AT&T's horizontal and vertical integration were anti-competitive. Facing a likely court defeat, AT&T executives preferred voluntary horizontal divestiture of local operations to increased regulation and vertical divestiture of Western Electric. The resulting 1982 Modification of Final Judgment divested AT&T's monopoly local service but allowed it to compete in long distance services, equipment sales and new areas. The existing 22 local operating companies were divided into seven groups, the only configuration that would allow approximately equal size. The resulting regional holding companies (or "Baby Bells") became seven separate companies on January 1, 1984 (Tunstall, 1985; Temin, 1987).

AT&T and its offspring eventually began to compete for the same service customers. As the Baby Bells became increasingly reluctant to buy equipment from their competitor, AT&T undertook a second, entirely voluntary breakup. In September 1995, it announced that AT&T Technologies would be spun off, and on Oct. 1, 1996 the new Lucent Technologies became fully independent, incorporating the former Western Electric and Bell Laboratories. Descended from the single AT&T in 1983, today these companies — AT&T, Lucent and the surviving Baby Bells — are central players in U.S. (and global) cellular telephone industry.[1]

Resource Supply: FCC's Spectrum Allocation Policies

Providing mobile telephone service differs from other telecommunications services in one key way: it requires the exclusive allocation of radio frequency spectrum.[2] In the U.S., this has been performed by the Federal Communications Commission, established by the Communications Act of 1934 and responsible for "establishing policies to govern interstate and international communications by radio, television, wire, satellite, and cable" (FCC, 1997: 14).

As an independent Federal regulation agency, the FCC's commissioners are appointed by the president but it must report to Congress. As such, the FCC is always at least indirectly considering competing interest groups — and directly if such groups have made themselves heard through a specific congressional mandate. In this regard, U.S. telecommunications policy prior to the 1980s was very different from most of Europe and Japan, where government PTT departments traditionally had both policy-making and operational responsibilities — which proved to be a crucial difference in the deployment of initial cellular telephone systems. The particular nature of the FCC, its procedures, and its permeability to a wide range of competing political influences explains why the U.S. was the first nation to invent cellular telephony but possibly as late as the 10th nation to commercially deploy it.[3]

Any mobile radio service depends on the exclusive allocation of frequency spectrum by the appropriate government agency. The availability of mobile telephony in the U.S. was inarguably delayed in the 1950s and 1960s due to limited spectrum allocated by the FCC. Unfortunately for mobile telephone operators and users, their spectrum requests usually lost as part of zero-sum lobbying game with one of Washington's most powerful lobbies, television broadcasters. Mobile telephone operators were also divided, between conventional wireline telephone companies (AT&T and smaller operators) and the numerous radio common carriers (RCCs). The RCCs were small (often undercapitalized) local firms licensed by the FCC beginning in the 1950s to provide public and private mobile radio services, which were also allowed to provide limited mobile telephone service in competition with AT&T.

Repeatedly, AT&T applied for VHF and UHF spectrum for mobile telephone service in the range 10-1,000 MHz (Table 1). From 1945 until the mid-1960s, the FCC usually decided in favor of television broadcasters, approving only 54 channels split between wireline carriers and the RCCs (Young, 1979: 6). Even with short and infrequent telephone calls, because each frequency could support only one call at a time AT&T's 23-channel service for metropolitan New York was limited to 543 paying customers with a waiting list of 3,700 potential users (Calhoun, 1988: 31). Even with increased spectrum, it was clear that such mobile telephone systems would be inadequate in major metropolitan areas. Therefore, AT&T had based its long-term mobile telephony plans since 1947 on the assumption it would use cellular radio.

The preferences received by a small number of broadcasters over land mobile use (public safety, private commercial services and common carriers) were challenged during the mid-1960s, as land mobile users reached 2.3 million transmitters in 1965 (Telecommunication Science Panel, 1966: 13).[4] In 1965-1966, both by the

Date	Action
1946	AT&T begins (non-cellular) commercial mobile telephone service in St. Louis
1947	AT&T proposes a 150-channel mobile telephone system using 40 MHz
1949	AT&T proposes a UHF mobile telephone system, but the FCC allocates spectrum for television broadcasts
1958	AT&T proposes a 75 MHz system at 800 MHz
1962	AT&T demonstrates for FCC a test UHF cellular system in Murray Hill, NJ
Oct. 1966	Commerce advisory panel report questions FCC allocation priorities
1968	Congressional hearings on "crisis in land mobile communications"
July 1968	*Notice of Inquiry and Notice of Proposed Rulemaking* (14 FCC 2d 311) Proposes to reallocate UHF channels 70-83 for mobile radio use
May 1970	*First Report and Second Notice of Inquiry* (35 FR 8644) FCC allocates 75 MHz for wireline cellular telephone carrier
Dec. 1971	AT&T, RCA and Motorola file proposals to use 800 MHz band for cellular mobile telephone systems
May 1974	*Second Report and Order* (46 FCC 2d 752) FCC allocates 40 MHz per market for a single wireline cellular telephone carrier
March 1975	*Memorandum Opinion and Order on Reconsideration* (51 FCC 2d 945) FCC opens cellular licensing to any qualified common carrier
July 1975	Illinois Bell applies for permission to build Chicago development system
Feb. 1977	American Radio Telephone System applies for permission to build Washington/Baltimore development system
March 1977	FCC authorizes Illinois Bell's Chicago development system (63 FCC 2d 655)
Oct. 1977	FCC authorizes ARTS' Washington development system (66 FCC 2d 481)
Nov. 1979	*Notice of Inquiry and Notice of Proposed Rulemaking* (78 FCC 2d 984) The FCC begins process of setting policies for building and operating cellular telephone systems
April 1981	*Report and Order* (86 FCC 2d 469) FCC adopts rules for cellular applicants, providing for two carriers (a wireline and a non-wireline) to each operate a 20 MHz system
Feb. 1982	*Memorandum Opinion and Order on Reconsideration* (89 FCC 2d 58) FCC reaffirms application procedures, except AT&T is only wireline required to maintain a separate cellular subsidiary
July 1982	*Memorandum Opinion and Order on Further Reconsideration* (90 FCC 2d 571-582)
June 1982	FCC accepts applications for two licenses in each of 30 largest metropolitan markets
1983	FCC grants first commercial cellular licenses
Oct. 13, 1983	Ameritech Mobile Communications (an AT&T subsidiary) launches the nation's first commercial cellular system in Chicago
1986	FCC increases cellular spectrum allocation from 30 Mhz (666 channels) to 40 Mhz (832 channels)

Table 1: U.S. mobile telephone regulatory milestones, 1946-1986

FCC and a government advisory board advocated reform in spectrum allocation policies. In 1968 a House of Representatives committee noted that mobile communications received only 4% of the spectrum below 960 MHz (1% for mobile telephony), vs. 87% for broadcasters (Calhoun, 1988: 48).

In response to escalating political pressure, in 1968 the FCC opened related policy-making inquiries, one of which (Docket 18262) became the basis for eventual U.S. cellular telephone service. Not surprisingly, the case brought intense lobbying on the one side by mobile radio operators, manufacturers and users, and on the other side by broadcasters. In 1970, the FCC tentatively allocated 75 MHz in the range 806-947 MHz to cellular mobile telephony and dispatch services, which Calhoun (1988: 49) terms "the second great watershed (after the invention of FM) in the development of mobile telephony."

Based on detailed technical plans for cellular systems submitted by Bell Laboratories and Motorola, in 1974 the FCC reduced the spectrum for cellular telephony to 40 MHz, but, as in the 1970 plan, assumed a single cellular system in each market to be operated by the local telephone company. A 1975 revision allowed applications by any qualified entity, including RCCs. Although the cellular license was not guaranteed to the local wireline carrier, at this point the FCC still felt, as Carson (1979: 320) put it, that "the expense, spectrum, requirement and wide coverage of a mature cellular system dictated that only one system in each urban area would be feasible." AT&T appealed the 1974 spectrum reduction, but it based its plans (as did the FCC) on the assumption that the full 40 MHz would be available to a single carrier.

Two teams applied for permits to build experimental ("development") cellular systems and were granted permission by the FCC in 1977; these were the later only two systems to officially come online in 1983, the first year of official U.S. cellular service (Seybold & Samples, 1986). A local AT&T operating company, Illinois Bell Telephone Co., applied for a license to build a system in Chicago, using equipment designed by Bell Laboratories and built by Western Electric (Huff, 1979); a radio common carrier, American Radio Telephone System, asked permission to operate a system in the Washington/Baltimore area to be designed and built by Motorola (Mikulski, 1986).

While AT&T had been planning cellular telephone systems for years, its role in radio system design was limited by two factors. First, by surrendering rights to manufacture and operate private mobile radio systems in the 1956 Consent Decree, it had only minimal radio expertise compared to leading U.S. maker, Motorola (Calhoun, 1988). Secondly, the FCC allowed Western Electric to manufacture base station equipment but not mobile equipment.[5] Therefore, when Illinois Bell and Bell Labs began validation of the Chicago system with an "equipment test " (July-December 1978), the test used 135 car-mounted mobile telephones manufactured by Oki Electric of Japan using a combined AT&T/Oki design. The second phase — service test with nearly 2,000 customers that began on December 12, 1978 — used leased telephones manufactured by Oki, Motorola and E.F. Johnson (a maker of 2-way mobile radios for RCCs) (Fisher, 1979; Huff, 1979; Blecher, 1980). Motorola, in fact, developed four generations of portable cellular phones before the 1983 commercial launch, with each generation smaller and lighter than its predecessor

(Lynn et al., 1996).

Based on quarterly operations reports from both the AT&T and Motorola systems, the FCC began regulatory proceedings that set the rules for operating cellular systems. Not only did the rulings define the competitive landscape until the mid-1990s, but they also are the clearest explication of U.S. policy in cellular communications for that period. The first ruling noted the changes in U.S. (wireline) telecommunications policy since it began examining cellular telephone systems, specifically the liberalization in private lines and terminal equipment. It also acknowledged the concerns of a Federal appeals court that AT&T would ultimately "operate most, if not all, of the cellular systems put into operation " (FCC, 1980: 987).

In light of these new developments, in 1981 the FCC proposed that the 40 MHz of spectrum be split between multiple applicants in each market, eventually allotting 20 MHz to two operators per market: one license reserved for an existing ("wireline") telephone in the market, another for a non-wireline carriers.[6] Subsequent rulings rejected nearly all challenges to the plans by AT&T, other telephone companies, RCCs and equipment makers.

The FCC gave the top priority to 306 metropolitan markets which accounted for 77% of the U.S. population; the rest of the country was divided into 428 rural service areas. Except for the most isolated rural markets, two bands were to be licensed for each market (Paetsch, 1993; Cellular Telephone Industry Association, 1996). In 1982, the FCC began by accepting applications for the 30 most populous markets. A flood of non-wireline applications were driven by "gold rush" forecasts of cellular wealth; by the June deadline, the FCC received 138 A block (non-wireline) and 52 B block (wireline) applications (Dizard, 1982; FCC, 1983: 62).

For the 30 B block licenses, 12 were uncontested; the remaining 40 applications (by AT&T, GTE and small independent local telephone companies) were quickly resolved by a series of joint venture agreements for the remaining 18 markets. With all 30 wireline license awards thus uncontested, rivals worried that the B block operators would be quickly licensed while the A block applicants were delayed by fights at the FCC and the courts (Dizard, 1982). These fears were realized: in 24 of the markets, the B block carrier was online first (by an average of 18 months); in only four markets was the A block carrier first (Seybold & Samples, 1986).

GTE gained control of the B block license in 7 markets; AT&T gained the remaining 23, but these were divided among the seven regional holding companies in the 1984 AT&T breakup. Among the radio carriers, no company controlled more than three licenses, although two firms (MCI and McCaw) had at least minority stake in six licenses (Table 2).

For the 30 largest markets, the FCC awarded contested licenses based on detailed business and technical plans, and accepted applications for the next 60 markets based on similar rules. With the difficulty choosing the most qualified applicant — and the certainty of a legal challenge by the losing party — it became quickly apparent that awarding licenses in all 306 MSA markets would take many years (except where competing applicants negotiated their own solution). Instead, the FCC awarded only the 30 largest markets based on competitive applications, but used a lottery of existing applicants for the next 60 markets. This brought a flood of speculative applications — 96,000 for markets 91-306, as compared to 1,200 for

Company	First Market	Date	Markets†	Largest Market	Remarks
AT&T Spin-oiffs					
Ameritech	Chicago	Oct. 83	4 (1)	Chicago (3)	Acquisition by Southwestern Bell proposed, 1998
Bell South	Miami	May 84	3	Miami (12)	
NYNEX	Buffalo	Apr. 84	3	New York (1)	Merged with Bell Atlantic 1997
Bell Atlantic	Washington	Apr. 84	4 (1)	Philadelphia (4)	
US West (New Vector)	Minneapolis	June 84	4 (1)	Minneapolis (15)	Wireless assets acquired by AirTouch, 1998
Pacific Telesis	Los Angeles	June 84	2 (2)	Los Angeles (2)	Wireless spun-off as AirTouch Communications, 1993; acquired by Vodafone (U.K.), 1999
Southwestern Bell	St. Louis	July 84	3 (1)	Dallas (9)	
Independent Wireline					
Centel			(2)		Acquired by Sprint, 1993
Contel			(2)		Acquired by GTE, 1991
GTE	Indianapolis	May 84	7 (5)	San Francisco (7)	Acquisition by Bell Atlantic proposed, 1998
United Telephone			(2)		Wireless assets acquired by Centel, 1988
Radio Carriers					
Cellular Communications			(4)		Entered 50/50 JV with Pacific Telesis, 1990
Communications Industries (Gencom)	San Diego	Apr. 86	1 (2)	San Diego (18)	Acquired by Pacific Telesis, 1986
Graphic Scanning			(2)		Acquired by Bell South, 1990
LIN Broadcasting	Philadelphia	Feb. 86	2 (2)	Philadelphia (4)	Control acquired by McCaw Cellular, 1990
McCaw Cellular	Kansas City	Feb. 86	1 (5)	Kansas City (24)	Acquired by AT&T, 1994
MCI AirSignal			3 (3)	Pittsburgh (13)	Assets acquired by McCaw, 1986
Metro Mobile CTS	Phoenix	Mar. 86	1 (4)	Phoenix (26)	Acquired by Bell Atlantic, 1992
Metromedia	Chicago	Jan. 85	1 (3)	Chicago (3)	Acquired by Southwestern Bell, 1986
Mobile Communications Corp. of America			(1)		Acquired by Bell South, 1989

† Majority owned (minority share) licenses
Note: Top 30 markets based on 1980 census
Source:Adapted from Seybold & Samples (1986); excludes Los Angeles A block license

Table 2: Original wireline and non-wireline AMPS licensees in top 30 markets Notes

markets 1-90; the lottery was also used for the rural service areas (Paetsch, 1993: 152-153).

Coupled with the AT&T breakup, the FCC achieved one of its primary goals of assuring that no single carrier dominated cellular service nationwide. The price was a complex system of technical and contractual solutions that were needed to allow users to roam between states (or even neighboring cities), delaying such mobility for at least five years. The FCC allocation schemes also gave a clear head start to the incumbent telephone companies as many non-wireline applicants fought for each A block license.

Experience in Other Regions

The three major sources of technological innovation in the early cellular telephone industry were North America, Japan and Europe (principally Northern Europe).[7] These three research centers were constrained by identical laws of physics and similar availability of enabling technologies such as microprocessors. In addition, competitive intelligence transferred technical knowledge between rival firms — primarily from AT&T to its challengers through the mid-1970s, but in many more directions as the Japanese, Nordic and rival US makers challenged AT&T's technical monopoly.

So if the determinants of technological diffusion were primarily technical, we would expect convergence of outcomes between these three pioneering regions (if not other laggards such as France and Germany). However, differing outcomes — particularly the speedy deployment of cellular systems outside North America — demonstrate the effect of differing policies which effectively delayed introduction of cellular service in the U.S. behind the other two regions.

Japan. As in the U.S. the central issue for the development of mobile telephony in Japan was the changing role of the dominant wireline telecommunications provider. However, unlike the U.S., that service provider began cellular service as a government PTT. Telephone service was provided by a succession of government ministries from 1889 until after World War II. In 1952, the Diet transferred telephone responsibilities to Nippon Telegraph and Telephone (NTT), a newly-created government corporation with a monopoly over domestic telephone sales and service, while the new Kokusai Denshin Denwa (KDD) was awarded a monopoly on international service. The NTT budget was approved annually by the Diet, and NTT workers were considered civil servants and thus denied the right to strike (Vogel, 1996: 139-140).

This structure was retained through the 1980s, when the liberalization (and breakup of AT&T) in the U.S. prompted Japanese regulators, telecommunications executives and users to study increased competition in the domestic market. The end result was that in April 1985, NTT became a quasi-private non-monopoly provider under the regulation of the Ministry of Posts and Telecommunications (MPT), although the majority of its shares continue to be government owned (Vogel, 1996; West et al., 1997). Vogel (1996) notes that while liberalization in Japan meant an increase in competition, it did not mean deregulation: outside North America, liberalization has meant a shift from PTT government departments (with both regulatory and service responsibilities) to government regulation of (partially or wholly) privatized telephone companies.

While NTT was horizontally integrated and contained its own research laboratories, unlike AT&T it relied on outside manufacturers (the "den-den" companies) to produce its equipment (West et al., 1997). NTT researchers developed their own analog cellular standard in the 1970s and began operating in Tokyo in December 1979 — the world's first commercial cellular system. By 1985, its 43,000 subscribers were primarily in the Tokyo and Osaka regions (Kuramoto & Shinji, 1986).

While analog licenses were allocated in each market between the monopoly wireline carrier and one rival, the similarity to the U.S. ends there. NTT held a nationwide monopoly in local services, lacked competition in pre-cellular mobile service and began its cellular service long before the existence of rival telecommunications companies. NTT was also effectively unregulated: the small size of MPT's telecommunications bureau prior to 1985 meant that NTT made its own autonomous policy decisions with little oversight (Johnson, 1989; Vogel, 1996).

Japan's 1985 telecommunications liberalization brought competition both to wireline and mobile services. Two other systems were granted licenses to compete with NTT's mobile services (MPT, 1996):

- Nippon Idô Tsûshin (IDO) began service in Tokyo in December 1988 and was also licensed for the Nagoya region; it is controlled by Toyota Motors.
- The "Cellular" group of companies, which began with Kansai Cellular Telephone in July 1989, and by 1990 covered seven of nine regions — all except the two served by IDO. Each local company is approximately 60% owned by DDI, in turn is a partly-owned subsidiary of Kyocera Corporation .

IDO used NTT's second-generation, higher capacity analog system while the DDI companies employed a variant of the Anglo-American TACS promoted by Motorola (Tyson, 1993).

Europe. As in the other developed countries, for Europe the mid 1980s marked a gradual liberalization of competition in telecommunications: as in Japan, the PTT's gradually shifted from government departments to public corporations, or even independent firms. Most European countries, however, lagged the U.S. and Japan in telecommunications liberalization and the introduction of cellular service.

Two major multi-country standards quickly gained most of Europe's analog cellular subscribers. Nordic Mobile Telephone (primarily in Northern Europe, Netherlands and Switzerland) and Total Access Communications System (U.K., Ireland, Spain and Italy) each held about 40% share of European subscribers in 1991, with proprietary systems in Germany, France and Italy accounting for most of the rest. The four Nordic countries (Sweden, Finland, Norway and Denmark) held 75% of the total subscribers in 1985, but this dropped to 31% in 1990, with the more populous United Kingdom accounting for 33% of the total (Paetsch, 1993: 280-283). These two regions — Nordic countries and the U.K. — were the most innovative in the first decade of European cellular systems.

Like the U.S., the Nordic countries possessed vast unpopulated stretches where a car-mounted telephone provided the only reliable communications technology. To cope with this demand, Swedish Telecom introduced three noncellular different mobile telephone systems: MTA (1956), MTB (1967) and MTD (1971).

Meanwhile, the Nordic PTT's, seeking the economies of scale possible from a common pan-Nordic system, evolved the MTC system towards a microelectronic-based cellular system which became NMT (Hultán & Mölleryd, 1995). Sweden introduced the first NMT system in the 450 MHz band in October 1981, followed by Norway, Denmark and Finland. Unexpectedly high demand prompted the four countries to deploy the NMT technology in a new 900 Mhz band in December 1986 (Paetsch, 1993).

The NMT-450 and NMT-900 systems were exported to several other European countries, promoted by PTT's and Nordic manufacturers seeking export sales. Historical path dependencies had already brought limited telecommunications deregulation to Sweden and Finland:

- In Sweden, telecom equipment for domestic use was manufactured by a joint venture of the PTT and private firm Ericsson, which derived the rest of its revenues from exports. Meanwhile, Swedish Telecom encouraged manufacturers and resellers to market mobile telephones because it lacked the capital to finance them (Noam, 1992; Hultán & Mölleryd, 1995).[8]
- In Finland, wireline service territories were divided between the PTT (Telecom Finland) and a coalition of 51 smaller local operators. Its leading telecom manufacturer, Nokia (and its Mobira radio subsidiary) faced strong competition from Ericsson in its home market and thus was also forced to concentrate on exports (Noam, 1992).

However, what liberalization that existed did not extend to arms-length licensing of frequency spectrum use by the PTT. For example, Sweden did not fully separate telecommunications and radio spectrum regulation from Telecom Sweden until July 1993 (Hultán & Mölleryd, 1995).

Meanwhile, the innovations in the U.K. were regulatory rather than technical. Lagging the research efforts in the U.S., Japan and the Nordic countries, and lacking domestic firms strong in mobile radio, British regulators instead selected a slight modification of the off-the-shelf American AMPS system (Taylor, 1985). By doing so, their analog systems enjoyed far greater success than those countries (e.g., Germany and France) which domestically developed their own cellular technology rather than employ proven technology from one of the three major design centers.

However, the U.K. led Europe and most of the world in competitive innovations with its aggressive liberalization and regulation of the previously monopoly British Telecom.[9] From the beginning, it licensed two competing systems, Cellnet and Vodafone which both began service in January 1985; it also limited BT to a 50% share of the former system. It continued to be Europe's most competitive market in the early 1990s with seven licensed carriers.

European technical, producer and regulatory institutions converged with the introduction of a second-generation, digital cellular system, GSM. Development of a pan-European standard was begun in 1982, and was transferred in 1989 to the new European Telecommunications Standards Institute, which represented countries in the European Community as well as non-EC countries such as the four NMT sponsors. First deployed in 1991, GSM uses a common standard and 900 MHz spectrum allocation throughout the European Union, allowing customers to roam between countries. Most national governments agreed to license two competing systems in each country (Besen, 1990; Cheeseman, 1991; Paetsch, 1993).

Subsequent Standardization Efforts

Japan and the U.S. trailed Europe in their efforts to develop digital cellular standards. By being the first digital service, and focusing early on a multi-vendor, multi-country solution, the GSM developed the largest market share in third country markets.

The most direct competition between rival standards was seen in the new 1.9 GHz PCS spectrum licensed in the U.S. beginning in 1995. As with analog service, the FCC issued licenses on a market-by-market basis, but unlike AMPS, a total of six licenses were auctioned and no firm was assured incumbent "wireline" status. Also unlike the AMPS period, the FCC adopted a market-oriented standards policy for the new PCS bands, authorizing multiple standards. Operations quickly settled upon three standards: GSM and two digital extensions to AMPS. In early 1999, the two most compatible standards (GSM and IS-136) agreed to develop mutual gateways to allow roaming by their respective customers (Luna, 1999).

Beyond this competition in various U.S. markets, the rivalry between digital standards emerged as a full-fledged battle with the 1998 efforts to develop a so-called "third generation" (3G) global wireless standard. Two rival alliances — aligned with the GSM standard led by Ericsson and the CDMA standard developed by Qualcomm — fought to win support for rival proposals. With European firms seeking a single standard and key U.S. firms pushing for multiple standards, the 3G standards battle brought threats of patent lawsuits and trade sanctions, which were only resolved through a March 1999 cross-licensing agreement and asset sale between Ericsson and Qualcomm.

Such 3G efforts typify calls for "harmonization" of standards, where "harmonization" is defined by proponents as agreement upon a single, monopoly standard. Despite such frequent calls, standards rivalries continue in many aspects of mobile telephony. Competition between air interface standards has been extended into outer space with new low-earth orbiting satellite telephone systems such as Iridium and Globalstar. Rival standards continue to be promoted for transmitting data via mobile telephones. Finally, as firms view to hasten the convergence of cellular telephones and handheld computers, alliances have formed around rival cellular telephone operating systems such as Windows CE, Palm OS, Geoworks, and the planned Symbian joint venture.

IMPLICATIONS AND CONCLUSIONS

Evolving Spectrum Resource Allocation Policies

A combination of different policies were used by national regulatory bodies to allocate the scarce spectrum resource for cellular mobile telephony. They include:

- *No policy.* In most early cases where the PTT was a government corporation, little or no formal regulation existed and the PTT/operator decided what spectrum it needed, and gained all the spectrum for its own use by default. Even where regulation exists, in most countries the incumbent wireline carrier was assured a cellular license (or, in many cases, for each successive frequency band of cellular service).

*   *Competitive allocation.* Based on some combination of technical, economic or political merit, this was the policy used for allocating spectrum in the 30 largest markets for 800 MHz service in the U.S., and also for most licenses in Japan and the U.K. However, as the FCC discovered, it can be a very costly and time-consuming process, particularly if separate subnational licenses are awarded and if the political system permits appeals of the regulatory award (Taylor, 1985; Calhoun, 1988).

*   *Collaboration and collusion between applicants.* In some cases, the government delegated resolution of competing claims to the applicants themselves. In the 800 MHz U.S. cellular licenses, the FCC encouraged vendors to pool applications for individual markets to minimize the contested applications — which resulted in a hodgepodge of fractional ownership shares that differed for each market (Seybold & Samples, 1986). Similarly, in Japan the MPT encouraged the weakest applicants for the final 1,900 MHz PHS nationwide system to consolidate into what became the Astel group.[10]

*   *Random allocation.* When the FCC bogged down in its technical allocation of 800 MHz licenses, it switched to a lottery award (with post-selection verification of minimal qualification criteria). Such an approach assures a flood of unqualified applicants and prevents any bidder from achieving economies of scale.[11]

*   *Market allocation.* The use of a public auction had been considered by the FCC for allocating the original 800 MHz spectrum (cf. FCC, 1980: 1001). Such market mechanisms have theoretical advantages of fairness and efficiency, although these advantages had never been tested on such a scale. Critics — such as foreign operators and regulators — pointed out a key disadvantage: raising the cost of service to operators, which theoretically would be passed onto consumers. Other, more serious disadvantages would not become obvious until after the PCS auctions were completed — most notably, the bankruptcies of entrepreneurial bidders who failed to obtain financing after overbidding (Cramton, 1996; Congressional Budget Office, 1997).

The choice of policy is influenced by the priority given competing goals — efficient use of spectrum, high diffusion, low consumer prices and equity between license applicants. But the selection process is also dependent upon the path-dependent institutional context for a given country — both in terms of the transparency of political institutions to rival applicants, but also in terms of the incumbent roles of various potential applicants. For example, the RCCs — independent radio carriers predating liberalization of the 1980s — have no real analog outside the U.S. and the U.K. Similarly, only in the U.S. and Finland did independent wireline incumbents play a significant role in challenging the dominant carrier's monopoly.

The allocation of radio spectrum (e.g. for broadcasting) has traditionally included regulatory specification of the communication standard to be used for that spectrum. That was the pattern for the NMT, AMPS and TACS analog services, as well as the digital services in Europe and Japan; as with broadcast standards, the mobile telephone standards were selected based on proposals by the prospective operators and/or manufacturers. However, trade friction in Japan and a modified *laissez faire* standards policy in the U.S. provided operators with a choice of standards; current trends in 3G standardization suggest that such intramarket standards competition may have been a temporary aberration rather than a trend for the future.

Competition and Consolidation

Operators. In determining competition policies for cellular telephone service, regulators measured market power (or its inverse, competitiveness) in two ways: at the national level (i.e., the share of the overall market), and in each individual market (i.e., the number of competing choices available to a potential subscriber). As a consequence of antitrust concerns related to AT&T, U.S. policies emphasized both forms of competition. In most other countries, only the latter competition was emphasized and licenses were issued nationwide.[12] Even in Japan, the 58 licenses issued for nine local regions were effective controlled by six groups which operated a total of nine systems.

The FCC licensing policies resulted in the initial fragmentation of U.S. licenses. For the wireline licenses, the divestiture of cellular licenses to the Baby Bells prevented what would have been an AT&T-led oligopoly (in concert with GTE and various rural telephone companies); instead, the seven "Baby Bells" started with roughly similar cellular assets (Noda, 1996). On the non-wireline side, the plethora of entrepreneurs and the FCC allocation process assured that the initial industry structure was even more fragmented than the eight major wireline carriers.

This fragmentation was gradually reversed through a process of mergers and acquisitions from 1985 to 1996. The mergers were sanctioned by the FCC, the Justice Department and Federal Judge Harold Greene, who approved all modifications to the 1982 AT&T anti-trust settlement.[13] The only merger limitation was that no firm could acquire a stake in both licenses in a local market, forcing some firms to sell or trade specific licenses before an acquisition. The result was that the largest initial carrier (GTE) had by 1996 dropped to fourth place. The largest carrier was AT&T, serving about one-fourth of the country, after having bought its way back into the cellular market through the acquisition McCaw Cellular — a new radio common carrier that had grown by acquiring other RCC cellular operators.

As cellular systems were deployed, most developed countries sought to encourage competition of the second type, allowing each consumer a choice between multiple cellular systems. Liberalization in wireline telecommunications encouraged competition in mobile telephony — which, in turn, encouraged further wireline liberalization. This helped erode the century-old concept of telephone service as a "natural monopoly" (Friedlander, 1995; Vogel, 1996).

The U.S., Canadian and British cellular systems were designed from their initial deployment to be duopolies, a pattern followed by second-generation systems in many (but not most) other countries. Successive generations of cellular technologies gave regulators the opportunity to increase competition beyond the duopolies as new spectrum was granted to a combination of new and existing entrants. For example, the FCC awarded six new licenses for 1,900 MHz Personal

Communications Service for each market, based on three successive auctions from 1994 to 1997 (Cramton, 1996; Congressional Budget Office, 1997). With 800 Mhz and mobile radio operator Nextel, a U.S. market had a maximum of nine licensed cellular systems, compared to seven in Japan and the U.K. In each of these countries, regulators promoted new entrants on the belief that increased competition spurred price cuts and diffusion of service.

Despite the lead of these three countries in promoting cellular competition, the highest market penetration rates through 1996 were achieved in the Nordic countries.[14] Except for Sweden, these countries began analog cellular service with government PTT monopolies, and were much slower in expanding the number of operators to the levels seen in the U.S., Japan or the U.K. Roos (1993) contends that policies based on competition between private operators inherently *reduce* diffusion of the technology — arguing instead that strong state-owned monopolies in the Nordic welfare states facilitated the universal deployment of cellular infrastructure, and thus, Northern Europe's exceptional cellular telephone diffusion rates. Knuutitila (1996) attributes successful demand-push diffusion to a high level of cultural cohesion built through shared understanding. Between countries, Nordic PTT representatives countries had worked closely together for many years; within countries, the PTT operator, private manufacturers and government had a long history of cooperation.

Manufacturers. The industry structure of manufacturing has changed surprisingly little during the 18 years since the launch of the first cellular systems. This stability is a testament both to the high barriers to entry in telecommunications manufacturing, as with the estimated $300 million AT&T and Motorola paid to develop cellular systems prior to system launch (Taylor, 1985). But it is also a testament to the permanence of core competencies based on tacit knowledge which shape the acquisition and production of new knowledge (Cohen & Levinthal, 1990).

Throughout the period, the leading makers of mobile terminals (sold to individual users) remain Motorola and Nokia — which aggressively introduced handheld portables to AMPS and NMT respectively as their corresponding rivals (AT&T and Ericsson) focused on car-mounted telephones. While Japanese makers as a group made significant inroads into the mobile terminal market (utilizing proprietary miniaturization and battery technologies), no one firm achieved a global presence to match Motorola, Nokia and Ericsson.

The market for cellular infrastructure — sold to system operators — consisted of two major components, mobile telephone switches and radio base stations. Not surprisingly, the leading infrastructure vendors corresponded to the major makers of digital switches for wireline telephony when cellular began: AT&T (now Lucent) and Ericsson. AT&T of course began as the largest maker of telephone equipment in the world, and launched its cellular products with the largest share of the largest market. Meanwhile, Ericsson proved to be the only firm to successfully combine both halves of the infrastructure puzzle, aided by its purchase of radio pioneer Svenska Radioaktiebolaget (Hultán & Mölleryd, 1995). The cellular infrastructure product lines for other vendors were strong only in their original, pre-cellular competency — either telephone switches (AT&T, Northern Telecom, NEC) or radio equipment (Motorola, Nokia).

Conclusions

The initial deployment of cellular telephone service followed the pattern of technology push often seen in technology-driven industries. Unlike many of these industries, the government possessed effective technology policy instruments to facilitate or delay deployment of cellular services. In particular, the inherent requirement for exclusive allocation of scarce radio frequency spectrum gave the government absolute control over the number and timing of new market entrants. When contrasted with other countries that pioneered cellular technologies, three specific path dependencies help explain the timing and eventual industry structure for cellular telephone service in the United States.

First, policies for mobile telephone entry were heavily influenced (if not determined) by changing ideology of telecommunications policy — specifically, the role of the near-monopoly national ("wireline") telecommunications carrier, AT&T. While other countries (notably Japan and the U.K.) emulated U.S. liberalization in wireline services (Vogel, 1996), the unique complexity of U.S. mobile telephone regulation and industry structure depended on the specific path taken by AT&T — both in its original vertical and horizontal integration, and its subsequent 1984 and 1996 breakups.

Secondly, while technological limitations prevented the introduction of cellular telephone service before the mid-1970s, the time required by the Federal Communications Commission to establish cellular policies delayed deployment of cellular telephone service in the U.S. by at least five years to its eventual 1983 start (Calhoun, 1988). In almost every country that launched its cellular system before October 1983, the operator was a government PTT that possessed the *de facto* (if not *de jure)* authority to determine its own policies for spectrum allocation and market entry.

Finally, two groups of competing claimants for spectrum allocation exploited the transparency of the pluralistic FCC regulatory policies to delay its allocation of mobile telephone spectrum between 1946 and 1983. During the initial 25 years, established television broadcasters prevented the reallocation of under-utilized spectrum from broadcasting to mobile communications, until a shift in Congressional policies increased the salience of the latter. After that, incumbents in the mobile radio industry (both RCCs and manufacturers like Motorola) fought every attempt by AT&T to obtain a wireline monopoly for cellular telephone service, accounting for most of the delays during the 1970s. The final delays came as potential applicants for cellular licenses (including AT&T and the RCCs) fought to obtain advantage in the FCCs eventual licensing policies.

The moderate success of cellular deployment in the U.S. despite these liabilities is a testament to the familiar American advantages of advanced technological capabilities and a large domestic market (Chandler, 1990). The diffusion of cellular telephones to U.S. consumers — and the major role played by U.S. manufacturers and operators in the global market — also stand as testament to the fundamental competence of the FCC and private firms despite the complexities of U.S. telecommunications policy regimes.

ENDNOTES

1   In 1997, the SBC Communications/Pacific Telesis and Bell Atlantic/NYNEX mergers reduced the Baby Bells to five; mergers pending as of this writing would reduce the number to four.

2   Spectrum allocation is also essential to other services such as pagers, cellular data and satellite phones, but here I focus on cellular telephony, the largest mobile service in terms of revenues and spectrum allocation.

3   The first system was in Japan (1979), followed by Saudi Arabia, Sweden, Norway (1981), Denmark, Spain (1982) and Finland (1983); less clear are Canada and Italy (Betteridge & Pulford, 1981; Paetsch, 1993; Meurling & Jeans, 1994). A detailed investigation of all systems would be necessary to determine whether each start date is more comparable to the Chicago service test with 1,900 paying customers (1978-1983) or its first year (1983-1984) of "official" service with 13,000 customers (Blecher, 1980; Cooper, 1985).

4   Most of these were mobile two-way radios; as of December 1977, there were still only 143,000 mobile telephone units in service (Young, 1979: 6).

5   A subsequent order (86 FCC 2d: 497-498) allowed AT&T to make mobile terminals if they were sold by an AT&T subsidary separate from the local cellular service provider — a plan rendered moot by the AT&T break-up. However, this policy shift came after the launch of AT&T's Chicago development system.

6   The 1981 ruling specified 666 channels using a total of 40 MHz in the range 825-890 MHz; in 1986, the FCC expanded this to 50 MHz and 832 channels.

7   Canada adopted the U.S. technology, radio frequencies and duopoly approach. The exception was the pioneering Aurora system in Alberta, which began service in 1982 before the U.S., but like Sweden's Comvik, quickly became a technological dead-end (Betteridge & Pulford, 1981; Mehrotra, 1994).

8   AT&T similarly encouraged rival radio manufacturers to compete in the production of mobile telephones — not because of lack of capital, but because the 1956 Consent Decree prohibited it from making mobile radios (Calhoun, 1988: 51).

9   Vogel (1996) notes that liberalization does not always translate to deregulation. In fact, outside the U.S. and Canada, liberalization has meant a shift from PTT government departments (with both regulatory and service responsibilities) to government regulation of (partially or wholly) privatized telephone companies.

10 Interview with Noriko Karaki, Deputy Director, Land Mobile Communications Division, Ministry of Posts and Telecommunications, Dec. 18, 1996.

11 However, such economies can be achieved through subsequent mergers and acquisitions, as occured in the U.S. throughout the late 1980's.

12 One must allow for a certain amount of geographic determinism. Fragmenting licenses by region spread the capital costs for infrastructure development in the U.S. and Canada, but most European countries were too small in terms of population or geography to justify such fragmentation.

13 Greene's oversight was ended by passage of the Telecommunications Reform

218 West

Act of 1996, which returned primary control of telecommunications policy to the FCC.

14 Diffusion rates in Norway and Denmark were surpassed by Iceland, a fifth Nordic country that did not participate in the NMT development but adopted the NMT and successive systems.

REFERENCES

Antonelli, Cristiano. (1994). Localized technological change and the evolution of standards as economic institutions. *Information Economics & Policy,* 6 (3-4), 195-216.

Arthur, W. Brian. (1989). Competing Technologies, Increasing Returns, and Lock-In by Historical Events. *Economic Journal,* 99, 116-131.

Berg, Sanford V. (1989). The Production of Compatibility - Technical Standards as Collective Goods. *Kyklos,* 42, (3), 361-383.

Besen, Stanley M. (1990). The European Telecommunications Standards Institute: A Preliminary Analysis. *Telecommunications Policy,* 14 (6), 521-530.

Besen, Stanley, M. and Joseph Farrell (1991). The Role of the ITU in Standardization: Pre-Eminence, Impotence or Rubber Stamp? *Telecommunications Policy,* 15 (4), 311-321.

Betteridge, R.I. and J. Pulford (1981). The Aurora (Automatic Roaming Radio) cellular AMTS system. *Global Communications,* 3 (4), 40-8.

Blecher, Franklin H. (1980). Advanced Mobile Phone Service. *IEEE Transactions on Vehicular Technology,* 29 (3), 238-244.

Calhoun, George. (1988). *Digital cellular radio.* Norwood, Mass: Artech House.

Carson, Viginia. (1979). Historical, Regulatory, and Litigatory Background of the FCC Docket No. 18262, 'An Inquiry Relative to the Future Use of the Frequency Band 806-960 MHz'. *IEEE Transactions on Vehicular Technology,* 28 (4), 315-326.

Chandler, Alfred D., Jr. (1977). *The visible hand: the managerial revolution in American business.* Cambridge, Mass.: Belknap Press.

Chandler, Alfred D., Jr. (1990). *Scale and scope: the dynamics of industrial capitalism.* Cambridge, Mass.: Belknap Press.

Cheeseman, David. (1991). The pan-European cellular mobile radio system. In R.C.V. Macario (Ed.), *Personal & mobile radio systems,* 270-289. London: P. Peregrinus.

Cohen, Wesley M. and Daniel A. Levinthal. (1990). Absorptive Capacity: A New Perspective on Learning and Innovation. *Administrative Science Quarterly,* 35 (1), 128-152.

Congressional Budget Office. (1997). *Where Do We Go From Here? The FCC Auctions and the Future of Radio Spectrum Management.* Washington, DC: Author.

Cooper, Martin. (1985, May). Cellular History: Made in Chicago. *Mobile Communications Business,* 32-34.

Cowan, Robin, Dominique Foray, and Georges Ferné. (1991). *Information technology standards: the economic dimension.* Paris: Organisation for Economic Co-operation and Development.

Cramton, Peter C. (1996). The PCS Spectrum Auctions: An Early Assessment. *Second Annual Conference of the Consortium for Research on Telecommunicatons*

Policy.

Cellular Telephone Industry Association. (1996, Spring). *The Wireless Marketbook.* Washington, DC: Author.

David, Paul A. and Mark Shurmer. (1996). Formal standards-setting for global telecommunications and information services. *Telecommunications Policy, 20* (10), 789-815.

Dizard, John W. (1982, July 12). Gold Rush at the FCC. *Fortune, 106,* 102-112.

Dosi, Giovanni. (1984). *Technical change and industrial transformation: the theory and an application to the semiconductor industry.* London: Macmillan.

Farrell, Joseph and Garth Saloner. (1988). Coordination Through Committees and Markets. *Rand Journal of Economics, 19* (2), 235-252.

FCC. (1980). An Inquiry Into the Use of the Bands 825-845 MHz and 870-890 MHz for Cellular Communication Systems: Notice of Inquiry and Notice of Proposed Rule Making. 78 FCC 2d, 984-1015.

FCC. (1983). 48th Annual Report/Fiscal Year 1982, Federal Communications Commission, Washington, D.C.: Author.

FCC. (1997). 62nd Annual Report FY1996, Federal Communications Commission, Washington, D.C.: Author, http://www.fcc.gov/Reports/ar96.pdf

Feldman, Steven P. (1986). *The Culture of Monopoly Management: An Interpretive Study in an American Utility.* New York: Garland Publishing.

Fisher, R.E. (1979). A subscriber set for the equipment test. *Bell System Technical Journal, 58* (1), 123-43.

Friedlander, Amy. (1995). *Natural monopoly and universal service: telephones and telegraphs in the U.S. communications infrastructure, 1837-1940.* Reston, Va.: Corporation for National Research Initiatives.

Funk, Jeffery L. (1996). World's Fastest Growing Mobile Communicatons Market — Japan. Asian Technology Information Program, Report 96.070.

Gabel, H. Landis. (1987). Open Standards in the European Computer Industry: The Case of X/Open" in H. Landis Gabel, ed., *Product standardization and competitive strategy.* Amsterdam: North-Holland.

Huff, D.L. (1979). The developmental system. *Bell System Technical Journal, 58* (1), 249-69.

Hultán, Staffan and Bengt C. Mölleryd. (1995). Mobile Telecommunications in Sweden. In Karl Ernst Schenk, Jürgen Müller and Thomas Schnöring (Eds.), *Mobile Telecommunications: Emerging European Markets.* Boston: Artech House.

Johnson, Chalmers. (1989). MITI, MPT and the Telecom Wars. In Chalmers Johnson, Laura D'Andrea Tyson and John Zysman (Eds.), *Politics and Productivity: the Real Story of Why Japan Works.* Cambridge, MA: Ballinger.

Kahaner, Larry. (1986). *On the line: the men of MCI – who took on AT&T, risked everything, and won!* New York, NY: Warner Books.

Kindleberger, Charles P. (1983). Standards as public collective and private goods. *Kyklos, 36* (3), 377-396.

Knuuttila, Juha. (1996). The ancient success regimes of ARP, NMT and GSM standards. Working paper, University of Jyväskylä, Department of Computer Science.

Kuramoto, Minoru & Masaaki Shinji. (1986). Second generation mobile radio telephone system in Japan. *IEEE Communications Magazine, 24* (2), 16-21.

Langlois, Richard. (1992). External economies and economic progress: The case of the microcomputer industry. *Business History Review,* 66 (1), 1-50.

Levitt, Theodore. (1960). Marketing Myopia. *Harvard Business Review.*

Lynn, Gary S., Joseph G. Morone and Albert S. Paulson. (1996). Marketing and Discontinuous Innovation: The Probe and Learn Process. *California Management Review,* 38 (3), 8-37.

Luna, Lynnette (1999, Feb. 15). UWCC and GSM aim for interoperability. *RCR,* 18 (7), p. 1,74.

Macario, R. C. V. (1997). *Cellular radio: principles and design* (2nd ed). New York: McGraw-Hill.

Mehrotra, Asha. (1994). *Cellular Radio: Analog and Digital Systems.* Boston: Artech House.

Meurling, John and Richard Jeans. (1994). *The mobile phone book: the invention of the mobile phone industry.* London: CommunicationsWeek International.

Mikulski, James J. (1986). DynaT*A*C cellular portable radiotelephone system experience in the US and the UK. *IEEE Communications Magazine,* 24 (2), 40-46.

Moore, Geoffrey A. (1995). *Inside the Tornado: Marketing Strategies from Silcon Valley's Cutting Edge.* New York: Harper Business.

Mowery, David and Nathan Rosenberg. (1979). The influence of market demand upon innovation: a critical review of some recent empirical studies. *Research Policy,* 8, 102-153.

MPT. (1996). *Idô Tsüshin Shistemu Guido 97* [Mobile Communications System Guide 97]. Tokyo: Ministry of Posts and Telecommunications, Land Mobile Commnications Division.

Noam, Eli. (1992). *Telecommunications in Europe.* New York: Oxford University Press.

Noda, Tomoyoshi. (1996). Intraorganizational strategy process and the evolution of intra -industry firm diversity: a comparative study of wireless communications business development in the seven Bell regional holding companies, unpublished D.B.A. dissertation, Harvard University, Graduate School of Business Administration.

Paetsch, Michael. (1993). *Mobile Communications in the US and Europe: Regulation, Technology, and Markets.* Boston, Mass.: Artech House.

Roos, J.P. (1993). 300,000 yuppies? Mobile telephones in Finland. *Telecommunications Policy,* 17 (6), 446-458.

Seybold, Andrew M. & Mel. A Samples. (1986). *Cellular Mobile Telephone Guide.* Indianapolis: Sams.

Taylor, J. (1985). The Government Viewpoint. In *Cellular radio in the UK: proceedings of a two-day conference held February 1985.* London: Oyez Scientific & Technical Services.

Telecommunication Science Panel of the Commerce Technical Advisory Board. (1966). Electromagnetic Spectrum Utilization — The Silent Crisis, Washington: Department of Commerce.

Temin, Peter. (1987). *The Fall of the Bell System,* Cambridge, Cambridge University Press.

Tunstall, W. Brooke. (1985). *Disconnecting Parties; Managing the Bell System Break-Up: An Inside View.* New York: McGraw-Hill, .

Tyson, Laura D'Andrea. (1993). *Who's bashing whom? trade conflicts in high-technology industries.* Washington, DC: Institute for International Economics.

Vogel, Steven. (1996). *Freer Markets, More Rules: The Paradoxical Politics of Regulatory Reform in the Advanced Industrial Countries.* Ithaca, NY: Cornell University Press.

West, Joel, Jason Dedrick and Kenneth L. Kraemer. (1997). Back to the Future: Japan's NII Plans. in Brian Kahin and Ernest Wilson (Eds.), *National Information Initiatives: Vision and Policy Design,* MIT Press, Cambridge, Mass.

Young, W.R. (1979). Advanced Mobile Phone Service: Introduction, background, and objectives. *Bell System Technical Journal,* 58 (1), 1-14.

Acknowledgments

The author wishes to acknowledge the valuable comments on earlier drafts of this article provided by Jason Dedrick, Jeffrey Funk, Kalle Lyytinen and John King.

Chapter XIV

How to Distribute a Cake Before Cutting it into Pieces: Alice in Wonderland or Radio Engineers' Gang in the Nordic Countries?

Vladis Fomin and Kalle Lyytinen
University of Jyväskylä

INTRODUCTION

This article analyses social networks by looking at the standard making processes. As a framework for analysis, actor network theory is chosen. Standards are of particular interest for actor network theory for they provide mechanisms to align interests of multiple social groups organized in networks that have a joint incentive in working with the standards and /or associated technologies. These social groups include scientific communities, government institutions and social movements (industrial groups, companies, and consumers) that are interested in regulating and innovating with new technologies. Standards provide the mechanisms to inscribe subsequent behaviors that are expected to become persistent over time.

Standard making process is a social process. Actors are involved in the process of continuous negotiation of their interests. Due to this fact, standards became an object of analysis for scholars within the social shaping of technology theory (SST). Though usually scholars of this school take standards as material objects, they interpret technology as such, e.g., a bicycle, or a steam machines. In Information Technology (IT), domain standards are intangible. Those are electronic data exchange formats, communications protocols, signalling protocols, etc. Wireless and mobile communications in particular, being a large field of IT, represent an interesting case for analysis. Present in mobile telephony's domain are *de jure* (e.g., GSM) and *de facto* standards (e.g., NMT). Also the broad scope and large scale of standardization processes suggests non-unified pattern of standard making and complex organizational structure. To make mobile telephony standards successful implies large networks and numerous mandatory passage points.

In this paper we apply actor network theory based analysis (ANT) to the development of NMT wireless standards. Researchers interested in IT standardization, except for a few studies on electronic data interchange (EDI) by Hanseth (1997), have overlooked this approach. The acronym NMT stands for Nordisk MobilTelefon (Nordic Mobile Telephone) and it can be historically regarded as one of the best examples of Nordic cooperation in technology as NMT systems have spread quite widely around the world and it also formed an important stepping stone for the evolution of GSM standards. We chose for ANT analysis of the NMT standard making process to learn of the usefulness of theoretical framework and to understand the standard making process of NMT as a social and institutional change. In our opinion, this more than anything else, explains the success of this interesting historical incident that changed the telecommunication industry radically and made Scandinavia a powerhouse of the wireless technologies. Our approach expected to bring more understanding on how the enthusiasm of a small number of actors fostered successful development of the NMT cellular telephony standard. At the same time the NMT standard was based on concepts and visions of its developers. Yet, it was these visions and engagements that lead to distributed the big cake of the cellular world even before cutting it into pieces.

The outline of the chapter is the following. In the next section, we discuss past theoretical analysis of the topic. Then we introduce new notions into ANT, such as a layer and a multilayered structure. Next we tell the story of the Nordic radio engineers' gang. We then analyze the NMT standard's development process as an instance of actor network mobilization. Some insights into future developments of cellular mobile communications, both from the technological and social perspectives are provided.

BACKGROUND

Technology is not a black box — this is not news to scholars of organizational processes (Callon, 1992; Lyytinen & Damsgaard, 1998; Pinch, 1988; Star, 1991; Williams, 1996). Any standard encompasses a body of knowledge and a social system (Pinch, 1988). Widely accepted Rogers' (1995) theory on diffusion of innovation was criticized as being inappropriate to account for complex technological innovations for it treats a technology as a material object lacking a social part (Lyytinen & Damsgaard, 1998). Thus an understanding of standard from inside, both as a body of knowledge and as a social system (Pinch, 1988) is needed in order to explain a success or failure of particular standardization process.

Cellular telephony, and NMT in particular, is a good example of complex, non-unified technology or its interpretative flexibility (Lyytinen & Damsgaard, 1998). The meaning of cellular technology under development was not the same for those involved. Operators aimed at longtime operation, profits were seen only in a long run (Meurling & Jeans, 1994; Myhre, 1998). Producers aimed at immediate profits and high volumes of production. Users aimed at the availability of service, its quality, high network coverage and affordability of the service. Regulatory bodies had to implement Pan-Nordic policies and resolve policy conflicts (Meurling & Jeans, 1994; Myhre, 1998; Toivola, 1992). What was in common for all four major groups of technology's users, is that they were foreseeing a new standard to be built on the existing telecommunications infrastructure. It would create a positive network externality — interconnections with the fixed telephone network would give an access to users of cellular phones, inherit its regulatory principles. When faced with diverse user groups in innovation process, alignment of multiple interests is required for the social construction of the innovation's significance, the negotiation of standards' scope and content, and the legitimization of the acceptable uses of the innovation (Lyytinen & Damsgaard, 1998).

We shall use the structure proposed by Lyytinen and Damsgaard (1998) to show the complexity of cellular technology, its infrastructural and networked nature. Cellular mobile telephony is inter-organizational in nature in that it involves operators, producers, regulatory bodies, R&D laboratories (not necessarily related to or established by producers), and users of cellular phones. Due to the fact that it is complex and innovative in nature, it requires know-how and skills to implement and operate. It relies on advanced existing telecommunications infrastructure, which creates a network externality factor and large set of dependencies to other components of the technological systems: fixed line switching, conventions on uses of phones, signalling protocols, service quality, etc. It is based on standards, thus creating a high degree of organizational interdependence. And last but not the least, it requires a large number of users to be efficient, thus pointing on importance of economies of scales. From these characteristics of the technology according to Lyytinen and Damsgaard (1998) the following conclusions can be drawn:

- The cellular technology involves the multiple path dependencies with earlier innovations (batteries, radio, etc.).
- Decision to develop and adopt the system is based on many simultaneous and not centralized development and adoption decisions.
- Success depends not on solely individual goals and desires, but also on the effectiveness of broader institutional and regulatory regimes. Presence of individual goals and desires vs. institutional and regulatory regimes suggests an approach of interwoven micro and macro analysis.
- It has high learning barriers due to its complexity.

It is not clear how the unit of analyses of the technology should be defined. Due to it is inter-organizational in nature, new forms of customer-supplier relationships can be implemented along all parts of the value system.

If standardization process of cellular telephony is to become successful, alignment of interests has to take place. With the number of actors and interests involved, when the alignment took place, the system becomes very stable, irreversible (Callon, 1992; Callon & Law, 1989), promising a "long life" to both the holders of the system, and its customers. For innovative technology in cellular business it takes not less than 10 years from inception to implementation. We learned it from experience with the first analogous system (NMT), followed by digital second generation's GSM system, and by the third generation public mobile telephony standard being currently under development. 10 years time was set for NMT by the Nordic Radioconference in the beginning of 1970s, and it took so long indeed (Meurling & Jeans, 1994; Toivola, 1992). Hans Myhre (1998) stated, "I think in fact that shift of basic technology has a ten years perspective". Though it should be mentioned that in the story of Nordic Mobile Telephone past dependencies could be traced back into the history of the social context: there were important earlier decisions on Nordic level what facilitated the decision making for NMT, which will be covered in the upcoming section.

THEORETICAL APPROACH

Pinch (1988) stated, "The starting point for an adequate understanding of any culture are the esoteric practices and bodies of knowledge which define that culture" (p. 70). To understand technology, or technology to be, there must be permanent communication channels established between those involved in knowledge acquisition and/or discovery. Large technological systems are both socially constructed and society shaping, they include physical artefacts and the organizations that use and manufacture them, as well as relevant legislative and regulative bodies and scientific communities (Lyytinen & Damsgaard, 1998). In order to bring stability to so largely diverse and complex network a standard is needed. NMT was the first successful commercial standard for cellular mobile telephony, by means of which diverse interests of the aforesaid broad network were aligned. Nonetheless, a standard itself is only an end product of highly complex standardization process, which requires analysis and understanding.

When studying organizations, there are two levels of analysis: micro and macro (Markus & Robey, 1988). Macro-level analysis implies studying of organization, environment, and social alliances. These entities are addressed as a whole. When there are decisions made individual preferences and goals of those people involved are not taken into consideration. In opposite, micro-level analysis treats decision making as a process determined by individuals. In the case of micro-level analysis decisions are determined by believes, goals, and interests of individuals. It is not the organization which makes decisions but individuals (Markus & Robey, 1988). When the research is based on only one of the aforesaid level analysis, the complete understanding of motivations and outcomes can not be obtained. Therefore it is helpful to address users and organizational context at different levels of granularity, by not only introducing a meso-level of analysis (what Markus and Robey (1988) define as mixed), but also by looking for dependencies by moving inside of the organizational-environmental structure from processes occurring on macro- to meso- and to micro-level and backwards. This kind of analysis should reveal the important passage points and the gatekeepers, associated with them. Delineating distinct hierarchical levels of social interaction will bring to definition of passage points and gatekeepers, named also as door keepers by Star (1991). The gatekeepers have a power to put forward or stop the standardization process (Star, 1991) and thus should be considered in analysis of standardization process

Actor network theory

One of the popular theories explaining dependencies of complex techno-social networks is actor network theory (ANT) (Callon, Latour, & Rip, 1986; Callon & Law, 1989; Latour, 1993). ANT studies the development of a scientific field and is identifying points of "interpretative flexibility". When applied to technological development, ANT seeks to identify points in time, where technology could be designed in more than one way, and to explain the dominance of one choice over another. ANT sees scientific and technical creation, as well as diffusion and consolidation of its results, stemming from interactions between actors. The actors are not necessarily humans — they could be well defined as technological artefacts. ANT concentrates on the analysis of possible choices made due to interactions between the actors (Callon, 1992): "How can we explain the fact that in certain cases, [technological] trajectories are successful and stabilize, whereas in others new configurations appear?" (p.72). Actor network theory studies the particular role played by technology, and the impact that technological artefacts have upon the socioeconomic context in which those exist (Callon, 1992): "the technical object is continually being reinserted into various socioeconomic contexts, which constitute different possible network configurations" (p.77). ANT intensively addresses a notion of alignment. For instance, if standardization process of cellular telephony is to become successful, alignment of interests of all involved has to take place. Alignment brings stability and irreversibility to the networks (Callon, 1992; Callon & Law, 1989), promising a "long life" to both the holders of the technological system, and to its customers.

Actor network theory is widely used to approach complex technological processes. Students of the domain are applying the theory to analyze the development and diffusion of material technical artefacts, such as, e.g., a bicycle or a steam machine (Callon, 1992; Star, 1991). In IT domain most of standards are intangible — protocols, data formats, signalling levels, etc. There are only few attempts to apply actor network theory to analysis of abstract, intangible standards. One of those rare examples is the work of (Hanseth, 1997) on electronic data interchange (EDI) standardization.

Applying ANT to the analysis of complex networked IT technologies, therefore, is a challenging task and requires introduction of several new perspectives to the theory. Uses of different levels of organizational analysis and treating standard as a boundary object we consider of substantial importance for more in-depth understanding of standard making process.

A notion of layer

In the Oxford English dictionary the meaning of the word "layer" is defined as "one who or that which lays (in various senses)". A layer of a process could be defined as a hierarchical level of organizational or environmental structure, where there are socio-technical processes taking place within that level. It could be also defined as a process on a certain hierarchical level of organizational structure. E.g., meetings of informal groups of a company are addresses as a micro-level process and are forming a "low" layer. When a meeting of the board of directors could be addressed as a meso-level process, and as a higher layer. Layers on the macro level are thus formed by inter-organizational processes, or in other words, by environment- or market-level processes.

It would be difficult to imagine complex organizational settings, where there was only one process-layer. As an organizational structure implies several hierarchical levels, there should be processes running on different levels of the organiza-

Macro Institutional environment:	Technological environment Technology	**Mixed** Plans:	Deliverables Organisation
Meso Strategy:	Monitoring Partners	**Micro** Micro behaviour:	Meetings Organisation Social Culture

Table 1. "Multiple layers of organisation"

tion, thus forming a multilayered structure (See Table 1).

Standard development processes can be seen to form layered structures, where different layers encompass varying social activities, from micro to macro, from small informal group meetings to strategy development. Stability and social order in these layers are continually negotiated as a social process of aligning interests. When dealing with networked social organization (Nohria & Eccles, 1992), where there is more than one institution involved, we come to the notion of a multilayered structure.

A notion of an actor of the layer

The Oxford English dictionary defines the actor as "1. One who acts, or performs action, or takes part in any affair; a doer. 2. One who personates a character, or acts." In this paper the notion of actor is taken from actor network theory scholars' works. According to Callon (1992), there are actors who recognize themselves in interaction. This interaction is embodied in the intermediaries that actors themselves put into circulation. Callon (1992) states, that an intermediary is anything which passes from one actor to another, and which constitutes the form and the substance of the relation set up between them — scientific articles, software, technological artefacts, instruments, contracts, money. E.g., if there is a PTT providing a one year budget for the NMT group, then the PTT is an actor attributing an intermediary — the budget.

According to Callon (1992) techno-economic network (TEN) is:

A coordinated set of heterogeneous actors — for instance, public laboratories, centres for technical research, companies, financial organizations,
users and the government — who participate collectively in the conception, development, production and distribution or diffusion of procedures for producing goods and services, some of which give rise to
market transactions. (p. 73)

Elements in such a network are not initially defined as human, social, or technological, by referred to by a common term — "actor". To understand the emergence and increase of TEN's that foster a development of standard and its delivery to the market, one should analyze heterogeneous activities that bring the aforesaid parties together. For that analysis we shall use concepts of intermediaries, actors, and translation.

In the given framework, an actor of the layer is a human or nonhuman having a certain role(s) in the process(es) taking place at that layer.

Enrolment of the actors

In the Oxford English dictionary the meaning of the verb "enrol" is defined as "to record, register; to include as a member, to record the admission of." In usual organizational settings actors of a social process are involved in interaction with other actors of the same process. A necessary condition for enrolling (engaging) an actor in organization process is aligning one's interests with the settings and requirements of the process. The "translation" will be defined as (Callon & Law, 1989): "a process in which sets of relations between projects, interests, goals, and naturally occurring entities — objects which might otherwise be quite separate from one another — are proposed and brought into being" (pp. 58-59). What is important about the translation, is that a successful translation will bring a durability to the network, it will align the diverse interests that different actors have, as Callon (1992) stated: "The successful translation creates the missed shared space, equivalence and commensurability: it aligns" (p. 84). From these definitions one could not see big ontological difference between Callon's "translation" and our "enrolment" — both are about aligning interests and interaction. Though we would argue that every time an actor is about to enrol to a certain social process, a mutual translation of interests must take place (See also Table 3).

A STORY OF THE NORDIC RADIO ENGINEERS' GANG

According to Williams (1996), "Standardization process involves economic and political processes in building alliances of interests with the necessary resources and technical expertise, around certain concepts or visions of a yet unrealized technologies" (p. 873). The notions of concept and vision appear in the Nordic Mobile Telephone system's development story. There was a group of radio engineers with their visions and skills. There was a concept of a new technology intended to be a part of the socioeconomic welfare system.

As per Meurling and Jeans (1994) 1993 was a crossover year, when mobile telephones became a part of a mass market and the consumer electronics industry, rather than the telecommunications industry. Weight, talk time and price drove the development of mobile handsets, and demand sometimes outstripped supply. Preconditions for this situation were set 20 years earlier, when the development of the first commercial pan-Nordic mobile system was initiated. In fact, in the early days most people thought that developing mobile telephony was simply a matter of exploiting radio technology, and solving radio-related problems. Myhre (1998) recalls, "Everyone thought it will be a radio system". There was also a market capacity underestimation (Meurling & Jeans, 1994; Myhre, 1998; Toivola, 1992). The huge number of subscribers for the future service could never be anticipated. Meurling and Jeans (1994) argue that market underestimation was also beneficial — it bought the industry time to plan, and to develop the technology in an orderly fashion, without pressure from eager money-makers-to-be. According to Myhre (1998), "Going back to 1969-70 at that time a time scale had to be rather long, because we had to do some research and lots of things. They said: 'OK, let's take away the pressure'".

Let's look at the initial stage of the NMT standard development. How was the idea given birth? Indeed, it would be rather problematic to identify the source of the idea. As was already mentioned, it was in the Nordic air for at least 10 years. But the inauguration of the idea has an official date and place. It took place in Kabelvåg, Norway, at the Nordic Teleconference in 1969. The decision was made to set up a work group whose task was, according to Toivola (1992), to investigate possibilities for creating a common traffic radio system for the Nordic countries. Carl-Gösta Åsdal, a representative of Swedish delegation reported to the delegates on the preliminary work that has been done in Sweden on the topic and drew conclusions of his study group on mobile telephony. Together with a couple of other delegates, he proposed that mobile telephony would be a highly worthwhile subject for Nordic cooperation. This proposal was immediately accepted, and the terms of reference for a joint Nordic working party were drafted at the same meeting. The first report from the Nordic working party, referred as NMT group, came in 1970. It recommended that a new, pan-Nordic automatic mobile telephone system should be developed (Meurling & Jeans, 1994; Toivola, 1992). Myhre (1998) stated, "So what they did is that there was established NMT group (or NTR 69-5) and they made a specification for a manual system in the 450 MHz band. And then they started to work [on] automatic system or what they thought should be an automatic system — they were not sure yet. The first activity was to find common frequencies".

As per Knuuttila (1997) there were slack resources that could be allocated for the NMT work, since the project was run by state owned monopolies of PTT's. Maybe a situation in which the numbers were modest, and hence the needs for funding relatively manageable, kept the decisions out of the boardrooms — people at operational levels were allowed to go ahead (Meurling & Jeans, 1994). This fact is also emphasized by Knuuttila (1997) and referred to as "radio engineers' gang". Almost all participants of the NMT development group belonged to the same engineers' generation and were able to communicate in "skandinaviska", the melange of Swedish, Danish and Norwegian. Myhre (1998) recalls, "They had this club and [they] talked the same language (almost)".

Similar discussions of developing the system were being held in AT&T in the US and also in Germany and France. The idea of developing such a system was there already before in the heads of Swedish and Finnish attendants of the conference, but this meeting provided an arena to discuss it and to form an organizational setting and institutional support for it (Calhoun, 1988; Meurling & Jeans, 1994; Myhre, 1998; Paetsch, 1993; Toivola, 1992).

Specifics of the process initiation

Speaking about preconditions of the development of the NMT standard one should consider the context in which the standardization process evolved.

The idea of creating mobile telephony system was in the Nordic air since the beginning of 1960s. Ten years before the first mobile system could be introduced (ARP), a need for this kind of system had been recognized. In Finland the matter was put forward by the director general of Finnish State Railways (VR), who in 1962 introduced the subject concerning the State Railways' line radio system, which was opened some time ago, to be used also as a telephone system for train passengers. It took 20 years to implement this idea to a successful business with no technical drawbacks. Nowadays, this kind of pay phone forms an important special sector of NMT. Their prototype was developed at VTT (State Technical Research Centre) and there are over 100 of them in trains, busses, and ships (Toivola, 1992). The first initiative leading to actual planning came from the Consultative Committee of the Field of Communications (CCFC). The minutes of the meeting 12.12.1966 read, as per Toivola (1992), like this, "1. The question of the necessity of a national radio phone network, keeping an eye mainly on road transport, was taken up for discussion... It was decided to establish a work division to investigate and clarify the question further..." (p. 20). The aforesaid shows that at the national level there were already in 1960s important decisions made, which had to pave the way for the future standards and services development.

On the Nordic level there were early envisages of possible ways of cooperation. In 1953, a permanent body called the Nordic Council was constituted by the governments of Denmark, Finland, Iceland, Norway and Sweden, to identify, study and recommend areas for cooperation between the member states across a wide variety of matters. Telecommunications was identified early on as an obvious area for Nordic cooperation (Meurling & Jeans 1994). Though any of the five Nordic countries was too small to develop a mobile telephone standard alone, the geography and the high level of telematic services in all Nordic countries, as well as "right-sizeness" became prerequisites to harmonize Nordic cooperation and develop the first trans-national mobile system.

Development and use of the Finnish national Auto Radio Phone (ARP) system have created the market and formed its channels (Knuuttila, Lyytinen, & King, 1996). It showed that there is a growing demand for mobile communications, it allocated human and material resources for a technical advance. Technology push and market pull forces were created and balanced over time of 20 years in Nordic countries. It is quite obvious that the initial idea of pan-Nordic mobile radio phone system was generated by "market pull" situation due to the ever-growing demand for ARP system. During the process of balancing market pull and technology push, highly motivated gangs of engineers acted as go-between among various Nordic PTTs' and industrial organizations (Knuuttila, Lyytinen, & King, 1996). The technological basis was there. A vision of market enlargement based on the growth experience of manually switched analogous ARP system was there. These reasons where enough to move ahead, to develop technological skills, and to create services and systems that could meet the growing demands (Knuuttila, 1997).

Not only economic factors fostered the Nordic cooperation. There were numerous cultural or ethnic factors to be considered. There was a vision of success among those involved. The deployment of Nordic traditions of cooperation as a given fact played an important role in making the NMT standard and the following technological success. The Nordic cooperation spirit gave every participating individual a motivation to present his own, his organization's and country's "voluntary" expertise to the standardization arena that was based on a free flow of innovations, inventions, contributions, and experiences. According to Myhre (1998), "We did not focus on that Dane should be there, or Finn should be there, etc. We put groups and the most competent guy was a chairman. And we were 5-6-7-8 people in each group. Different background, but all of them had interest in making these specifications".

Initial tasks and goals

What was so special about Nordic cellular system, that it became a story of success, when the others, much stronger markets, failed to implement the standard? Highlights from interview of Myhre (1998) are to provide the insight:

Q: We had discussion with non-Nordic colleagues. For them it is nearly impossible to understand this model of cooperation, that there were no lawyers, no accounting, and it was just doing this thing.

A: Yes, I think at that time when I came in at the late 1970s, we saw that they had a trial system in US. When you look at the map of America, you can see big lakes and there is a river going northeast — it was a trial system at that area. But they never lifted it up. They made a trial system, but they had problems of making it commercial. The driving element behind this, as they said, was serving customer. It was very much focusing on serving the customer. We had a manual system, and we had waiting lists for telephones here in Norway [...it was the same in Finland...] but it was driven by enthusiasm. And we did not look at the commercial side. Well, we did look, but it was not about the competition. We were cooperating, no fights, we had a support from the top management — it was driven by a common goal. And we saw that the only way to get this all to operation was by working together — we could not do it separately. It was too big project and the suppliers would not support us.

It becomes clear from the aforesaid, that the approach to the development process was different than that of US. We would argue that the enthusiasm of radio engineers' gang and the support from the top management of Nordic PTTs were crucial success factors. They were aiming at service, when the new-developing standard was intended to become a part of the socioeconomic welfare system of the Nordic Countries. The story of the success of the NMT, indeed, proves once again that technology is not a black box due to the fact that in US, Europe, and Japan they failed to bring the same technology to commercial success, when in the Nordic countries it was flourishing.

Initial specifications

What was an interesting aspect of initiation of the developing standard — the NMT — is a separation of service requirements from technical specifications. There were already requirements for quality and operational principles, to some extent, implied by fixed network. As Meurling and Jeans, 1994 stated: "Many conditions to be met by mobile technology are conditions set by fixed networks — the mobile network doesn't and can't have its own private set of rules" (p. 9). In February 1973 the NMT development group proposed a number of important planning principles in its report to the Nordic Teleconference held the same year. This document of 23 pages points out the guidelines for the development of the system without any technical solutions (Toivola, 1992). Only formulated ideas. Though that was sufficient, and when the technical specifications were finally developed, these were offered to international industrial marked free of charge. None of the NMT specifications and technical solutions were patented: the Nordic Administrations would be offering industry throughout the world a specification for an open international standard — free of charge (Meurling & Jeans, 1994).

TECHNOLOGY CONVERGENCE THROUGH ENROLMENT

This paper is about understanding the nature of stabilization of large-scale heterogeneous networks by means that include processes of standardization (Star, 1991). It is a standard what provides mechanisms to align interests of multiple social groups organized in networks that have a joint incentive in working with the standards and /or associated technologies. Standards themselves are imbedded into technical knowledge. According to Williams (1996), "Standardization process involves economic and political processes in building alliances of interests with the necessary resources and technical expertise, around certain concepts or visions of a yet unrealized technologies" (p. 873). Star (1991) stated, "It is through a series of translations that [one] is able to link very heterogeneous interests into a mini-empire, thus, in Latour's words, 'raising the world'... The enrolment does not just involve armies of people, but also of nature and technologies" (p. 28).

It becomes obvious, that 'raising the world' of emerging standard is about establishing communication channels, in order to delegate one's interests to others. This is about enrolling selves into other processes, other layers, markets. Myhre (1998) gives a good example of delegation of interests:

> With manufacturers we had different objectives. They've got shareholders, and if the manufacturer does not make profit, it will be decapitated by shareholders. While the mobile operator must think of a lot of other aspects: it has to make it as cheap as possible and attractive for the suppliers or its customers, but he can think on a long-term base system.

What becomes a crucial issue is finding an answer to the question: "Whom to enrol?" Whose interests should be aligned with interests of the standard's developer? To answer this question one should think of mandatory passage points of the standardization process. For the every mandatory passage point there should be a gatekeeper — an actor that has a power to make decisions within its layer. The initial point of 'raising the world' is the most difficult, for the initiator is faced with already stabilized networks, would those be manufacturers, regulators, etc. Thus at first, one should break the inner balance of the networks and reestablish them again. It does not sound to be an easy task. What is the right approach to do it? Star (1991) provides the answer to this question:

> But there is a question about where to begin and where to be based in our analysis of standards and technologies. Door-makers and keepers are good points of departure for our analysis, because they remind us that, indeed, it might have been otherwise (pp.52-53).

If we think of the second generation cellular telephony system's standard, like Nordic NMT or American AMPS, a non-exhaustive list of the passage points and

Passage points:	Gatekeepers:
frequency allocation	Regulator[1]
radio switching	standard developer
numbering	standard developer
signalling	standard developer
convincing manufacturers	standard developer
regulatory basis	regulator & operator(s)

Table 2. "The passage points and the gatekeepers of the NMT standard"

the gatekeepers could be defined as in Table 2.

A good example on how the enrolment of manufacturers took place can be traced in NMT development process. There was no ready technology in hand, e.g., for duplex radio filters. Lack of some technological solutions was partially a reason for a large time scale for the standardization process. The NMT group had to delegate to manufacturers its interests, but so that the proposal was attractive also for the manufacturer. Myhre (1998) gives the example of finding and enrolling gatekeepers:

We sent specifications to almost 200 addresses all over the world... Philosophy always was to communicate with industry, get their feedback, but we did the decisions...What we did later we sent specifications also to component manufacturers, because if we had sent specifications only to mobile manufacturers, they would think of it in a different way and make the solutions in different ways.

Infrastructure creates a soil which is either to accept or reject a new technology (Pinch, 1988). It was observed that technological expansion occurs when there is a match between a new techno-economic paradigm and the socio-institutional climate (Freeman, 1988). Levinthal (1998) defines two criteria which are determining the innovation development process:

The nature and pace of technological change are driven by two elements of the selection process. The first is a process of adaptation: the technology becomes adapted to the particular needs of the new niche that it is exploiting. The second element corresponds to the resource abundance of the niche. (pp. 220-221)

Though the existing infrastructure and its readiness to accept the new standard are necessary, those are not sufficient for the standard to become successfully adopted. Example can be taken from comparison of the Nordic NMT and American AMPS cellular systems' development and implementation processes. These two regions had a common infrastructural characteristics. There was a low threshold in getting a telephone (in US lower that in the Nordic region), there was high penetration of phone lines, in both regions operators had experience of managing large networks. Nonetheless, the Nordic cellular telephony system was implemented and received commercial success right after springing out of the laboratory, when the American system witnessed a delay in 10-15 years (Bekkers, 1998; Mehrotra, 1994; Paetsch, 1993).

Passage points:	Gatekeepers:	NMT	AMPS
radio switching	standard developer	success	success
numbering	standard developer	success	success
signalling	standard developer	success	success
convincing manufacturers	standard developer	success	success
regulatory basis	regulator & operator(s)	success	success
frequency allocation	NMT=operators AMPS=regulator	success	failure

Table 3. "Enrolling gatekeepers"

Knowing the history of cellular systems' development in both US and the Nordic Countries, and assuming that passage points and gatekeepers of the standardization processes which are given in *Table 3* are correct, if not exhaustive, we come to the conclusion, that not only the match with the existing infrastructure and abundance of resources (it was the case in both regions) determine the success of adoption of a novel technology, but also a social process of actors' enrolment. We come to this conclusion from approaching the frequency allocation failure in AMPS adoption process as a failure in negotiating interests between two social groups: the standard's developer (AT&T) and the regulators (FCC).

From micro to macro

Though it is not only a large institution, like FCC in the case of the AMPS standard, who can be a gatekeeper possessing the power needed to let the process go or to stop it. As likely a gatekeeper can be an individual. Toivola (1992) refers to an interesting case, when in order to avoid a mismatch of equipment produced by different manufacturers a method specifically for testing mobile stations had to be developed. A system simulator was needed which would emulate the functioning of the network. When Nordic work didn't seem to progress, the directors of Finnish PTA Radio Department got tired of bad reports and ordered the study from the Communication Laboratory of Helsinki University of Technology (TKK). Than a single individual, Master degree student Matti Ilmonen, became an actor capable of delivering the needed intermediary — a technical artefact, the simulator — which made it possible for the development process to proceed. This example makes us to realize the importance of the mixed level of analysis: moving from macro (institutional) to micro (individual) gives more clear insight in the analysis of complex socio-technical processes, like standardization process. Thus enrolment at different levels should be analyzed in order to obtain the most comprehensive picture of complex organizational process.

From enrolment to convergence

We can state that successful enrolment is a necessary factor for technology convergence. As technology encompasses both social and technological aspects, actor network theory suggests that enrolment is necessary for technological convergence. Convergence can take place only if there is a common ground for two

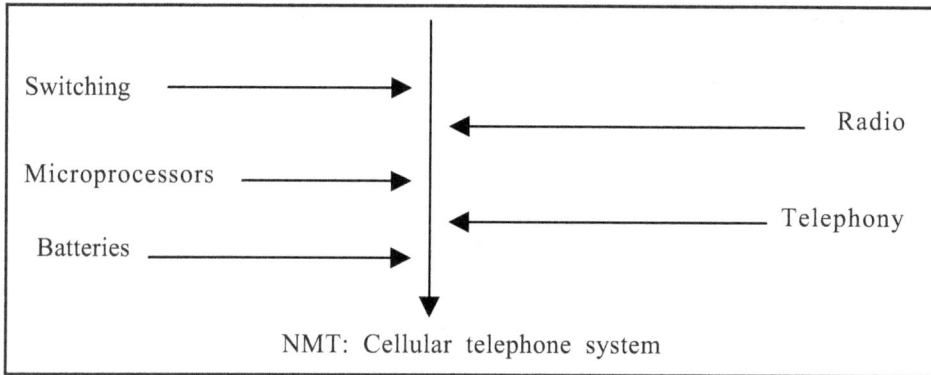

Switching ————————▶

Radio ◀———————————

Microprocessors ————————▶

Telephony ◀———————————

Batteries ————————▶

NMT: Cellular telephone system

Table 4. "Convergence of technologies"

distinct technologies, if the actors have negotiated and agreed on their interests in the novel technological paradigm. Since enrolment is about aligning interests, and according to Callon (1992) the successful translation creates the missed shared space, equivalence and commensurability — it aligns interests of the actors, then the convergence may take place.

In the case of cellular telephony we can observe a convergence of several technologies in a new distinctive one. These convergences can not necessarily be anticipated by the developers. An example of unexpected technological lineage was given Myhre (1998):

> When we went down in 1976 with a request for mobile switching, with a tender in fact, we got responses from Motorola, from company in Finland "Telefeno"... And then we are back to that balance between the telephone system and the radio system... Switching system with a radio part — this is what they realized when we've got rid of all these mental constrains.

Though convergence of switching and radio technologies was not the only one which was needed to have the second generation cellular mobile telephony standard in hand. There were numerous other convergences needed (See Table 4).

Standard as a boundary object

Alignment, enrolment, convergence — all the notions imply communications and exchange of information. Markus and Robey (1988) have shown the importance of communication in the high complexity process: "Unanalyzable, nonroutine tasks require rich information, capable of conveying complex and equivocal meanings; facial expressions and voice are the media most capable of processing this rich qualitative information" (p. 587). Thus we can argue that a standard can be "communicated" between actors involved in the standardization process. It becomes a mediator, an abstract artefact that aligns interests. Treating the standard as abstract representation of technical knowledge, as a social phenomena, knowledge creation and diffusion processes can be better understood, just as the standardization process as such (See Figure 1).

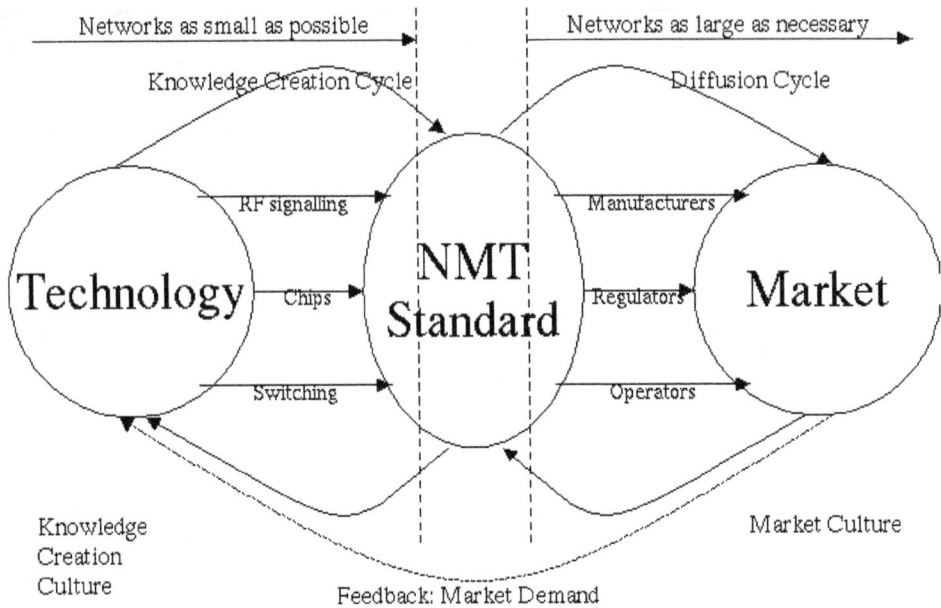

Figure 1. "Standard as a boundary object"

FUTURE PERSPECTIVES

As in the case of the NMT standard's development there was technology convergence observed, the following cellular mobile telephony standards have witnessed similar processes. If in the case of the NMT they needed to merge radio, switching, battery, phone, and microprocessor technologies, then for the GSM standard there were cryptography and digital radio added. The third generation cellular mobile system, UMTS, will utilize multimedia technologies, such as wireless video transmission. The primary communication media becomes data, not a speech, which was in the preceding two generations of cellular systems. The notion of speech becomes highly abstract, for it is interpret by the system as one of the possible instances of the data.

When novel technical paradigms emerge, depending on the impact of those paradigms on the existing ones, and depending on the quotient of change they bring, secondary novel technological systems may emerge (Freeman, 1988). In the case of the NMT it was a clearinghouse, which was a novel perspective in the mobile telephony business, and became a necessity due to the ever-growing number of mobile subscribers, who were roaming from one county to another. At one point PTTs, which were exchanging data tapes with accounting information, found themselves sunk into cross-calculations (Myhre, 1998). There had to be a more efficient data processing organizational unit, and it had to be outside the PTTs. Already with the NMT clearinghouse business was blooming, and it only strength-ened positions with emergence of the GSM standard. What kind of new domains we should expect from the introduction of the Third Generation of cellular telephony standard? Perhaps it will be specialized Internet services, audio/video software development for wireless devices, and likely, a shift in manufacturing from notebooks to powerful handheld mobile stations.

Along with changes in technological domains taking place within implementation of new standards, changes in social aspects of technologies can be identified. The interweave of multilayered relations of actors becomes more sophisticated due to increasing number of players. It changes the scope and the scale of the standardization process. Identifying necessary passage points becomes a more difficult task due to the broad technological scope. Increased number of actors widens the number of gatekeepers, and deepens diversity of interests. When the passage points are changing, comparing to those of the preceding generation of cellular telephony system, the gatekeepers are also changing, blurring the vision of future development of emerging novel standard.

CONCLUSION

Actor network theory is widely applied by scholars to analyze socio-technical processes. Though when analyzing technological development scholars focus on material standards, e.g. bicycles, steam machines, personal computers, etc. They study technology as such. In IT domain most of the standards are intangible — protocols, data formats, signalling levels, etc. There were only few attempts to apply actor network theory to analysis of abstract, intangible standards. One of those rare examples is work of Hanseth (1997) on electronic data interchange (EDI) standardization. Using actor network theory approach when dealing with abstract technological knowledge, when addressing standards as social processes, new notions should be introduced to the theory. We introduce notions of a layer and a multilayered structure in order to enable better understanding of socio-technical processes, which take place in a highly complex, networked, and infrastructural domain of IT. Using notions of layer and multilayered structure one can better understand how diverse interests of actors are negotiated and aligned, how actors from one organizational and/or hierarchical level are enrolled into processes on other levels. Multilayered organization structure emphasizes the importance of establishing communication channels between different layers and different players, which facilitates exchange of knowledge, and as a result, technology convergence and birth of novel technologies. Our approach brings more understanding on how an enthusiasm of actors at micro layer and a support of actors of macro layer fostered successful development of the NMT cellular telephony standard basing on only concepts and visions of its developers: how the big cake of cellular world was distributed before cutting it into pieces.

ENDNOTES

1. For the NMT it was a special administrative body subordinate to PTTs — frequency administration. They went to Nordic radio conferences and specified they were to allocate frequencies

REFERENCES

Bekkers, R. a. S., Jan. (1998). *Mobile Telecommunications: Standards, Regulation, and applications*. Boston, Loindon: The Artech House.

Calhoun, G. (1988). *Digital Cellular Radio*: The Artech House.

Callon, M. (1992). The Dynamics of Techno-Economic Networks. In P. S. Rod Coombs, Vivien Walsh (Ed.), *Technological Change and Company Strategies: Economic and sociological perspectives* (pp. 517): Harcourt Brace Jovanovich, Publishers.

Callon, M., Latour, B., & Rip, A. (Eds.). (1986). *Mapping the Dynamics of Science and Technology*. Basingstoke: Macmillan.

Callon, M., & Law, J. (1989). On the Construction of Sociotechnical Networks: Content and Context Revisited. *Knowledge and Society, 8*, 57-83.

Freeman, C. (1988). Induced Innovation, Diffusion of Innovations and Business Cycles. In B. Elliott (Ed.), *Technology and Social Process* . Edinburgh: Edinburgh University Press.

Hanseth, O., & Monteiro, E. (1997). Inscribing behavior in information infrastructure standards. *Accounting, management and information technologies, 7*, 183-211.

Knuuttila, J. (1997). *The Ancient Success Regimes of ARP, NMT and GSM Standards: The Rise and... of Chronocracy or Somersaults Backwards for the People*. Unpublished lecinciate thesis, University of Jyväskylä, Jyväskylä.

Knuuttila, J., Lyytinen, K., & King, J. L. (1996). The Parturition of Mobile Telephony Revisited: The case of Standardization and Institutional Intervention in the Nordic Countries : University of Jyväskylä, Dept. of Computer Science and Information Systems.

Latour, B. (1993). Ethnography of a "High-Tech" Case: About Aramis'. In P. Lemannier (Ed.), *Technological Choices* . London: Routledge and Kegan Paul.

Levinthal, D. A. (1998). The Slow Pace of Rapid Technological Change: Gradualism and Punctuation in Technological Change. *Industrial and Corporate Change, 7*(2), 217-247.

Lyytinen, K., & Damsgaard, J. (1998). *What's Wrong with the Diffusion of Innovation Theory? The case of Complex and Networked Technology* (Working paper). Jyväskylä, Finland: University of Jyväskylä.

Markus, M. L., & Robey, D. (1988). Information Technology and Organizational Change: Casual Structure in Theory and Research. *Management Science, 34*(5), 583-598.

Mehrotra, A. (1994). *Cellular Radio: Analog and digital systems*. Boston, London: The Artech House.

Meurling, J., & Jeans, R. (1994). *The Mobile Phone Book*. London: CommunicationsWeek International.

Myhre, H. (1998). *Story of the NMT system's development* [Interview]. Oslo.

Nohria, N., & Eccles, R. G. (Eds.). (1992). *Network and Organizations: Structure, Form, and Action*. Boston (Mass.): Harvard Bussines School Press.

Paetsch, M. (1993). *The evolution of Mobile Communications in the U.S. and Europe: Regulation, Technology, and Markets*. London: Artech House.

Pinch, T. (1988). Understanding Technology: Some Possible Implications of Work in the Sociology of Science. In B. Elliott (Ed.), *Technology and Social Process* . Edinburgh: Edinburgh University Press.

Rogers, E. M. (1995). *Diffusion of Innovations*. (4 ed.). New York: The Free Press. A

Division of Simon & Schuster Inc.

Star, S. L. (1991). Power, technologies and the phenomenology of conventions: on being allergic to onions. In J. Law (Ed.), *A Sociology of Monsters: Essays on Power, Technology and Domination* . London: Routledge.

Toivola, K. (1992). *Highlights of Mobile Communications development.* Helsinki: TELE.

Williams, R. (1996). The Social Shaping of Technology. *Elsevier*, 865-899.

Acknowledgments

The authors thank John Leslie King for his enthusiasm and sharing of ideas. We also acknowledge other members of STAMINA group for helping with interview materials and creating a fertile working atmosphere: Ari Manninen, Juha Knuuttila, Ping Gao, Anri Kivimäki, Joel West.

Chapter XV

The European Computer Driving Licence (ECDL): A Standard of Basic Competence for Personal Computer Users

Michael Sherwood-Smith
University College Dublin & ECDL Foundation

INTRODUCTION

There is a recognized need to spread computer literacy across every level of society to generate an inclusive global information society, in which every citizen has an opportunity to participate (WRC, 1998; Dolan, 1997; European Commission, 1996; Green Paper, 1996). The visionary comments and actions of Commissioner Bangemann of the European Commission sparked off the recognition of the need for a computer literate population. He suggested that launching initiatives in areas of education, training and work organization was a basic requirement for supporting inclusion for the citizen in the information society. Establishing the European Computer Driving Licence (ECDL) as a basic standard of computing competence for every citizen underpins this objective. Computer literacy programs based on encouraging and motivating people to obtain an ECDL support the objective of including all citizens in the development of our information society.

ECDL deployment programs have been launched to address the recognized need to increase computer literacy. Some of the issues raised in the European Commission reports were the following:

- Greater efforts must be made in our schools, to prepare the next generation to participate and benefit fully;
- Greater efforts must be made to stimulate European citizens to create content for new services whether education, entertainment or business;
- Continued efforts must be made to keep Europe at the forefront of technology

and infrastructure development and deployment for everyone;
- Sustained efforts must be made to increase the public awareness of the benefits of active participation in the information society;
- New collective efforts are needed to realize broader social benefits, particularly at local and community level.

The European Computer Driving Licence addresses most of the above issues. The overall objectives of the ECDL dissemination program are:
- To raise the level of computing competence for all European citizens, for those in the work force, seeking to join the work force, for those at home and for students.
- To increase the productivity of all employees who need to use the computer in their work.
- To enable better returns from investments in information and communications technology.
- To ensure all computer users understand the "Best Practices" and advantages of using a computer.

The European Computer Driving Licence (ECDL) is a certified standard of basic competence for the users of a personal computer (any brand of personal desktop or portable computer). The ECDL, discussed subsequently in detail in this Chapter, is a certificate awarded to a person who has achieved a basic standard of knowledge of the concepts of information and communications technology and has acquired a basic standard of competence using a personal computer.

The objective of this chapter is to describe the ECDL standard. It gives the background to what has been done in Europe, with the European Computer Driving Licence (ECDL) initiative. It describes the ECDL standard in terms of content, procedures for certification and the organization around the deployment of the ECDL. It concludes by outlining the development plans for the ECDL and the aspiration to establish a 'de facto' standard through the general acceptance and worldwide take up of the ECDL concept.

BACKGROUND AND OVERVIEW

The vision of a Computer Driving Licence (CDL) was first formulated in Finland in the early 1990s. The vision was realized in Finland with the introduction in January 1994 of the first Computer Driving Licence certification and deployment program. The program was support by the Ministry of Education, the Central Organization of Finnish Trade Unions, the Confederation of Finnish Industry and Employers, the Finnish Information Processing Association and the Ministry of Labour. The first Computer Driving Licences were awarded in early 1994 and well in excess of 45,000 have been issued.

The 'Computer Driving Licence' concept is based on the idea of a certified standard of basic competence for computer users. The driving licence establishes a standard for everyone who uses a computer in either a professional or a personal capacity. It is a certificate that verifies a user's competence, declares his or her computer skills and makes them readily mobile within the global information society and business world. The licence targets competence in, what might be termed, current office applications independent of the specific software or actual hardware platform used. Employers and job seekers all agree on the need for this standard definition of practical competence in information and communications technology. The CDL concept is realized around a *Syllabus* and a *Question and Test Base*. The users have to familiarize themselves with the syllabus and acquire the knowledge and practical skills outlined in the syllabus. To be awarded the licence users must pass standard tests to demonstrate their competence.

The concept of the Finnish CDL was taken up by CEPIS (The Council of European Professional Informatics Societies) who set up a 'User Skills Task Force' consisting of representatives from ten countries (Norway, Sweden, Denmark, Finland, Netherlands, Ireland, France, Austria, Italy and the UK) to consider the increase in competence required for the European work force. The task force looked for a suitable model and examined the Finnish CDL in detail. In the meantime, the Finnish program was developing successfully. It had awakened the interest of those who needed 'computing competence' and drawn attention to Information Technology (IT) and its widespread application in office work and in modern society. Companies did set up training programs for their employees.

After thorough study the CEPIS task force concluded that the basic Finnish concept was widely applicable throughout Europe, subject to some changes and updates. They followed their vision of a pan-European ECDL and committed themselves to further assess the changes required and the modifications needed to have the ECDL meet the requirements of a wider marketplace. This project was supported by the Commission of the European Union [Project No. ESPRIT 22561]. A series of pilot tests were carried out in Norway, Sweden, Denmark, France and Ireland. As a result of the tests and a thorough evaluation of the concept, modifications were agreed. A new *European Syllabus* and a *European Question and Test Base* were developed to meet the newly defined requirements. New structures, discussed later in this Chapter, were established to develop and deploy the ECDL.

The European Computer Driving Licence (ECDL) was formalized on the basis of the CDL. The ECDL certifies that a person has *knowledge of the basic concepts of information technology (IT)* and is *competent to use a personal computer* at a basic level of competence. In practice the ECDL certificate indicates that the holder has passed one theoretical test that assesses his or her knowledge of the basic concepts of information technology, and six practical tests - which assess the holder's basic competence in using a personal computer.

The domains covered by the ECDL are the following:

Theoretical:

Module 1: The Basic Concepts of Information Technology—the first basic requirement of competence in computing is to understand the context for computer-based applications in society and the key concepts of computers.

Practical:

Module 2: Using the Computer and Managing Files—it is important to understand the basic housekeeping functions required for the efficient use of the computer.

Module 3: Word processing—using the computer for the creation, editing, formatting, storing and printing of a document. Most documents used today are produced by word processing applications.

Module 4: Spreadsheet— this is similar to a manual spreadsheet, with the ability to perform calculations rapidly. It is used in preparing budgets, producing forecasts, business graphics and financial reports.

Module 5: Database— assists in the organization of large volumes of data to allow fast and flexible access to that data.

Module 6: Presentation—graphics have always been an important tool for architects, engineers, illustrators and designers. The use of computer-based presentation and drawing tools has grown in many application areas to support effective communication. These tools are used extensively in business and in teaching.

Module 7: Information and Communication—the use of networks has grown from a desire to share resources and to communicate with others. Today, millions of computers are connected together around the world. It is important that ECDL holders can make effective use of the "Information Super Highway".

As well as establishing the standard the CEPIS task force established the organization to deploy the ECDL. The first deployment program was launched in Sweden in 1996 and in 1997 the ECDL Foundation was incorporated. The organization for the deployment of the ECDL is discussed in the Section after next.

THE ECDL STANDARD (THE SYLLABUS)

The ECDL Syllabus

The ECDL Standard is based on the computer user knowing certain basic facts about Information Technology (IT) and having certain skills in using a personal computer and its software to carry out every day tasks using the functionality (capabilities) of the computer. The domain of knowledge covered and the sets of necessary skills to be mastered in the standard are described in the *ECDL Syllabus*.

The purpose of the ECDL Syllabus is to list the facts to be known and the tasks to be mastered that are covered by the ECDL standard. The ECDL Syllabus also expresses, in general terms, the level of knowledge and skill required to achieve the ECDL standard.

The Syllabus of Module 1 of the ECDL is described at a general level as a list of 'Knowledge Areas' these are detailed at a more granular level as 'Knowledge Items' [facts to be known] within the Knowledge Area. The Syllabus of Modules 2-7 of the ECDL is described at a general level as a list of 'Skills Sets' these are detailed at a more granular level as 'Task Items' [tasks to be mastered] within the Skill Set.

The ECDL Syllabus has the following uses.

1. To enable candidates to understand the Skill Sets and Knowledge Areas covered by the ECDL Syllabus.
2. To allow employers to understand and appreciate the skills and competencies of employees or job applicants to their organizations who have been awarded an ECDL.
3. To enable teachers, educators and courseware providers to develop curricula and training packages to cover the ECDL Syllabus.
4. To support the ECDL Foundation to approve curricula and training packages as covering the ECDL Syllabus.
5. To enable the ECDL Foundation to develop the European Question and Test

Base (EQTB) to measure a candidate's competence in a Module and certify that the candidate has reached the required level of competence.

The ECDL Syllabus is characterized as having a certain *stability*, consequently the Knowledge Items and Tasks are expressed with some generality so that the ECDL Syllabus does not have to be continually changed to track minor technical innovations of software products. The level of detail allows candidates, teachers and course providers to know what is covered by the Syllabus. Supplementary appendices may be provided with specific 'Question and Tests Bases' clarifying the inclusion or exclusion of some specific area of functionality or knowledge item in the 'Test Base'.

The context for the ECDL

The target population for the ECDL is the general public who wants to use a personal computer competently. For example, office workers who want their skill using a personal computer formally recognized, will want to take the ECDL tests and obtain certification. The formal certification is of value to employers to assess an employee or potential employee's skill. Workers, students and citizens who want to formally show with their certification that they have basic competence to use a computer, will want to take the ECDL tests and obtain an ECDL. The target population is a very broad population.

The ECDL is a certificate at a *basic* level of competence. This basic level of competence is expressed in the ECDL Syllabus by giving a detailed list of the Knowledge Items (Module 1) and Task Items (Module 2-7) covered by the ECDL Syllabus. The Task Items included in the Syllabus are those recognized by expert practitioners in the various domains as being necessary to cover basic competence. So the details of the ECDL Syllabus express the tasks to be mastered to acquire a basic level of computing competence. (Module 1: Knowledge of basic IT concepts).

The ECDL is a certificate of practical competence. The candidate for certification must have knowledge of, understand the purpose of and be able to perform the task in listed in the ECDL Syllabus. This implies that ECDL test questions can address: *knowledge* of facts concerning topics of the ECDL Syllabus, *understanding* of the purpose of the functional features of software for task items included in the ECDL Syllabus and the *ability to apply* the knowledge to deliver a task item outcome.

The Architecture of the ECDL Syllabus

The architecture of the ECDL Syllabus and its European Question and Test Base (EQTB) is shown in the diagram in Figure 1.

The ECDL Syllabus is a concise statement of the material covered by the ECDL. So strictly speaking the ECDL Syllabus (presented in part in Annex 1.) consists of Frame (A) and Frame (B). Frames (C) & (D) are included to complete the picture. Frame (A) represents, at a high level, the knowledge and skills, computer users with an ECDL, have acquired. Frame (A) is included as a separate frame to highlight the fact that the ECDL Syllabus is hierarchical and that the top of the hierarchy (Module/Category/Skill Set) is a summary in its own right, albeit at a high level, of the broad areas covered by the ECDL. A Syllabus at this level can be prepared and presented as a marketing document. Each Skill Set is then broken down into Task

(A) Identifiable Core of material covered by the ECDL Syllabus

(1) What the PC user needs to know.

(2) The skills and competencies offered to the employer.

ECDL Syllabus

Version 3.0

(C) Guidelines for ECDL certification

(1) How the PC user is going to be examined for ECDL certification.

(B) Detail of IT knowledge and user skills of the ECDL

(1) The detail of what the PC user needs to know and have mastered before presenting himself or herself for certification.

(2) What the PC user, teacher and trainer needs to have covered in ECDL training.

(D) The European Question and Test Base (EQTB)

(1) The world of questions from which the PC user is going to be examined.

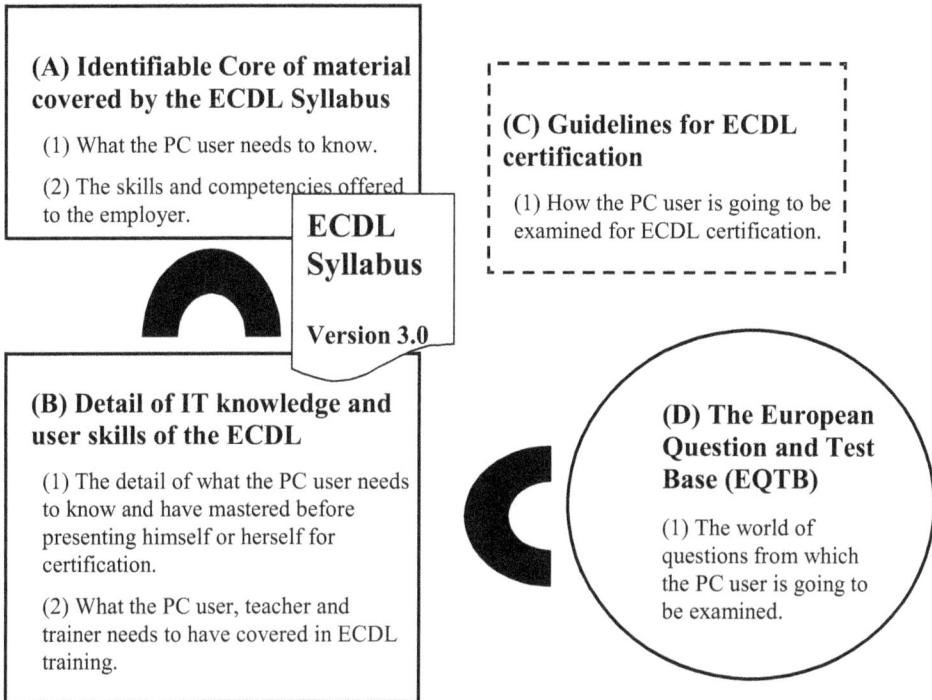

Figure 1: The architecture of the ECDL Syllabus

Items to express the detail of the ECDL Syllabus. Frame (B) represents a more granular description of what is outlined in Frame (A). The ECDL Syllabus at this granularity details precisely the IT concepts that must be known and the Task Items that must be mastered by the computer user to achieve the ECDL standard. The description is platform independent and software vendor independent. This detailed level ECDL Syllabus is provided for candidates, trainers and courseware providers. The whole of the hierarchy Frame A and B constitute the ECDL Syllabus Version 3.0.

Frame (C) represents the procedures of the ECDL Standard for user certification. These are not part of the ECDL Syllabus as such. The Guidelines for ECDL Certification are available as part of the Standards and Quality Guidelines for ECDL Operators. The Guidelines cover both manual and automatic testing. The underlying standard for manual and automatic testing is equivalent, but the detail of the guideline is different in its application to manual or automatic testing.

Frame (D) represents the EQTB. The ECDL may be certified by Manual Testing or Automatic Testing. The EQTB is the world of questions for testing the candidate's knowledge/ability to know/master the material of Frame (B). Currently there is a manual EQTB. An EQTB for automatic testing has been engineered for some test engines. The ECDL standard is vendor independent. The EQTB is the property of the ECDL Foundation and not in the public domain.

The Structure of the ECDL Syllabus
The ECDL Syllabus is structured as shown in the diagram below.

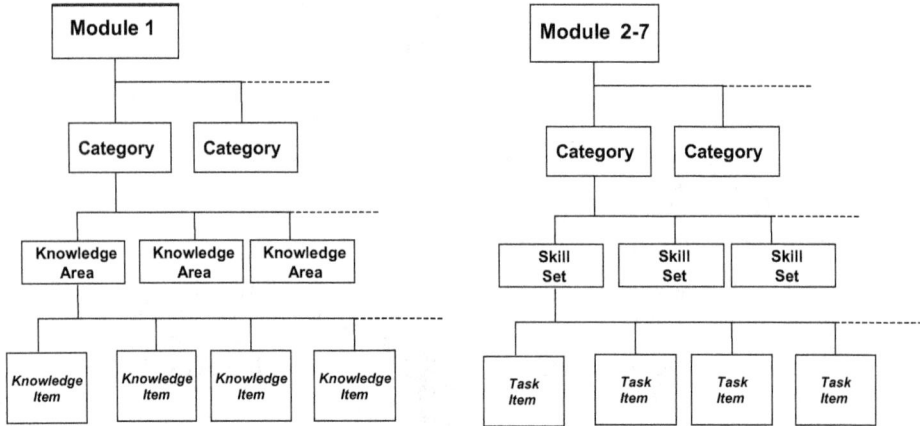

Figure 2: The ECDL Syllabus is structure

Module: The Module of the ECDL.

Category: The Category is a high level grouping of Skill Sets (Knowledge Areas) used to give a measured sequence to the development of the material in the ECDL Syllabus and help computer users and trainers to understand the progression in the material of the syllabus.

Knowledge Area: A label for a logical grouping of related IT facts.

Skill Set: A label for a logical grouping of Tasks to be mastered by the personal computer user to acquire that skill.

Knowledge Item/Task Item: (See Glossary). *Knowledge Item:* A fact to be known.

Task Item: A Task may be completed with a single action of the keyboard or a task may require several keyboard actions depending on the functionality of the software. The Syllabus expresses the Task Item in terms of the single outcome of the task (usually the Task Item requires only one keyboard action but sometimes a sequence of actions is implied). In the ECDL Syllabus every effort is made to keep a vendor independent description. Vendor Independence is sometimes difficult to achieve as 'de facto' computer literacy is associated with the most popular practices which sometimes reflect a vendor's particular implementation paradigm.

THE ECDL STANDARD (THE ECDL PRODUCT)

The ECDL product, for the user, consists of two physical items: The ECDL licence itself and the European Computer Skills Card (ECSC) which is an official record of the test passed so far.

The European Computer Driving Licence (ECDL) is a certificate that attests to the holder's ability to use a computer and common applications, and states that he or she has been tested according to the ECDL standards on the authorized ECDL Syllabus. The ECDL is an internationally accepted document. It will have a similar format throughout Europe. It will bear the name "European Computer Driving Licence" in English and also the name by which it is known in each individual country.

From the users point of view the gaining of a complete ECDL builds around the European Computer Skills Card (ECSC). This card records the skills demonstrated in each of the seven modules of the syllabus. Before taking the first test the candidate will buy a Skills Card or be given one by the sponsoring organization. An ECSC is an official form that is used to record the certification for each module test successfully completed. As the candidate passes each module successfully, the corresponding skill is noted in their card. The final achievement of a full card leads to the award of the licence. The candidate will send the ECSC to the country's national ECDL office where it will be exchanged for a European Computer Driving Licence.

The modules can be taken in any sequence and the tests can be taken in different test centres and indeed in different countries. An ECDL or ECSC that is granted in one country is valid in another. Both the ECDL and ECSC are internationally recognized certificates.

THE ECDL STANDARD (TESTING)

The current standard is based mostly on tests for manual testing with some automatic tests. The questions for the tests come from the European Question and Test Base (EQTB). The EQTB, which is not in the public domain, is the set of questions and tests which candidates must pass in order to attain an ECDL. An approved test centre selects questions and tasks for a specific test. The EQTB is based on the ECDL Syllabus consequently any change to the ECDL syllabus are reflected in new versions of the EQTB.

The EQTB contains a description of each module, and consists of:

The questions/exercises: For example, with manual testing, an exercise for Module 3 (Word Processing) might be to prepare the 'Agenda' for a meeting with particular layout specifications. Or for Module 4 (Spreadsheets) to prepare a Profit and Loss account given a certain set of figures. In automatic testing the questions look for simple keyboard actions or a sequence of keyboard actions.

Answer guidelines: For manual testing, the answer guide is an indication of model answers that should be accepted. The answer guide is quite extensive. The evaluation of the completed task is done with some degree of tolerance. For automatic testing a direct match for the answer is expected.

The skills required to pass the module are outlined in the ECDL Syllabus.

The marking guidelines: For manual testing these guidelines direct the examiner on the marks for each question and how to complete the prescribed standard Assessment Form for the module in question. For automatic testing the test engines follow a prescribed marking algorithm.

The actual format of the test between Module1 and Module 5-7 differ and the marking guidelines of the ECDL standard differ from Module to Module. For example Module 1 has a 60% pass mark and for some of the practical Modules the standard insists on correct answers for elementary (mandatory) tasks (90% pass mark) and a lower mark for other tasks. A high pass mark of 70-80 % is expected in general, reflecting the need to be competent in the practical skill being tested. With the current introduction of automatic testing the marking schemes for the ECDL are in full evolution. For the ECDL standard there is an established marking scheme for each Module.

The ECDL certification procedure is moving towards automatic testing. Automatic tests have been introduced in several European countries. The questions in the tests are multiple choice questions, a simulation of task-based questions using screen dumps of actual software and some test software implements 'in application' testing. Some test engines under evaluation by the ECDL Foundation are handling 'in application' questions (completely equivalent to the current manual testing). Evaluation and development of automated testing tools is a current development project.

DEPLOYMENT OF THE ECDL

ECDL Structures - The ECDL Foundation

CEPIS established the ECDL Foundation in January 1997. It is a coordinating, development and controlling body for the ECDL. Its role is described in more detail below. The ECDL Foundation is made up of CEPIS and a number of CEPIS member societies, who subscribe to the foundation by paying a membership fee. These are the Members of the ECDL Foundation. The current member countries are: [Austria, Cyprus, Czech Republic, Denmark, Estonia, France, Germany, Hungary, Ireland, Italy, The Netherlands, Norway, Poland, Portugal, Sweden, Switzerland and the United Kingdom] plus Finland, which has special status. It is in these 17 European countries that the ECDL is being deployed.

The ECDL concept has been internationalized as the International Computer Driving Licence (ICDL) with the same standard (Sherwood-Smith, 1998). There are Licensees in South Africa and Zimbabwe. In both those countries ICDL deployment programs have been launched based on the same standard as the ECDL. Australia and Canada are now launching programs and the international structure is being reviewed (discussed in the next Section).

The role of the ECDL Foundation has been clearly defined:

Product Development: The ECDL Foundation supported by its Member countries is responsible for: The review and development of the ECDL and skills card concept; the development of the ECDL Syllabus; the development of the ECDL Question and Test Base; supporting the design of automated test tools.

Quality Assurance: Establishing Standard and Quality Guidelines for ECDL processes to be used by Member organizations; Obtaining ISO certification of the processes; Controlling/Auditing conformance throughout participating countries.

Marketing: Signing up new ECDL Foundation members; Supporting the Member to develop their country; Information dissemination.

Financial Management: The ECDL Foundation is a 'not for profit' organization incorporated in Ireland. Income is generated mainly from royalties on the sale of European Computer Skills Cards.

The legal framework underpinning the ECDL Foundation is the following.

The rights of the Computer Driving Licence have been acquired by CEPIS. The full rights for Europe have been transferred from CEPIS to the ECDL Foundation following a royalty agreement. A further royalty agreement covers the transfer of the Worldwide rights to the ECDL Foundation.

ECDL Structures - Member Countries

The ECDL concept is owned by the ECDL Foundation. The ECDL rights, for a country, are usually transferred, under licence, to the CEPIS member organization, the professional Computer Society, in that country. The following diagram illustrates the context for ECDL dissemination throughout Europe.

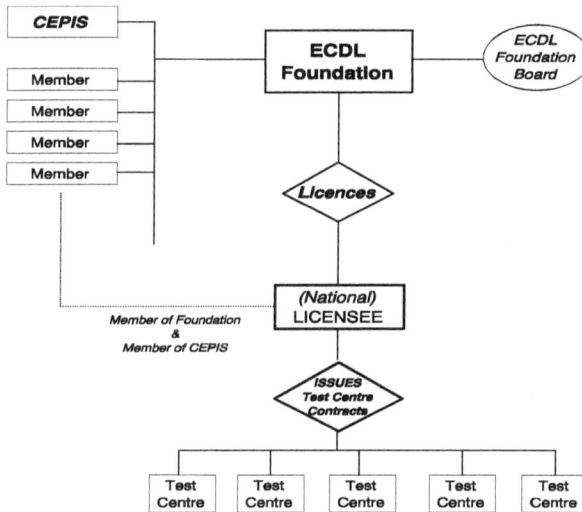

Figure 3: ECDL dissemination throughout Europe

The ECDL Foundation licences a (*National*) Licensee to use the ECDL Concept and to establish an ECDL dissemination program in that country. The (*National*) Licensee must qualify to be a member of CEPIS for European countries (Members of the EU, countries of Central and Eastern Europe and newly independent States of the former Soviet Union). This organization on paying its membership fee becomes a Member of the ECDL Foundation. Test Centres are set up in each country to administer ECDL tests. The ECDL Foundation will retains overall responsibility for ensuring that standards are met. Currently, outside Europe the ECDL Foundation licences organizations, qualified by the ECDL Foundation, as Licensees (under review).

The structure for evolving the ECDL Foundation to accommodate the concept of the ICDL is under discussion. The underlying ethos surrounding the ECDL concept and its establishment, as a standard, is that it has been promoted democratically and by the agreement of the ECDL Foundation Members. The Members are professional Informatics and Computer Societies in various countries. The requirements are driven by the market requirements, informed by recognized expert practitioners. It is through the support of Professional Societies across the world that the ICDL will most probably be developed and deployed.

A standard ECDL licence agreement governs the relationship between the ECDL Foundation and the Licensee, and prescribes the standards of operation required. The Licensee may operate the ECDL license themselves and enter into Test Centre contracts which entitles a Test Centre to use the ECDL concept and skills cards (ECSCs) in connection with the operation of Test Centres. Alternatively the Licensee may form some sort of partnership to set up an autonomous sublicense to operate the licence in that country. The Sublicensee may then enter into Test Centre Contracts to disseminate the ECDL.

ECDL 'Operator' is the term used for the organization controlling, managing and deploying the ECDL in a territory. By agreement, there may be several ECDL Operators in a national territory.

The ECDL Foundation ensures and controls that the heads of agreement of the standard contract are complied with in the operation of licences granted. One of the clauses of the standard contract is that there is an obligation on the part of ECDL Operators and Test Centre contractors to allow the ECDL Foundation, at all times, to have access for audit purposes. The auditing function is authorized to inspect, make copies of and extracts from all books, records, documentation, systems and software used and/or maintained by the Test Centre. The Standards and Quality Guidelines have been elaborated to help the ECDL Operators and Test Centres to operate in the agreed way.

The designated ECDL Operator for the territory develops the market. The marketing and promotion of the ICDL worldwide, ECDL across Europe and subsequently in a national territory is clearly a key topic. It is beyond the scope of this Chapter, which focuses on the *ECDL standard*, to fully discuss marketing. The heading is included here in recognition of the fact that the marketing of the ICDL and ECDL is a key activity of promoting and establishing the standard. The ECDL Foundation, as mentioned earlier, has an essential nurturing and driving role to play in the marketing of the ECDL and now the ICDL. European Union and Government aid may be sought to support the development of training programs, the marketing and deployment of the driving licence. An ECDL deployment program depends on the availability of suitable training courses leading to ECDL certification. The ECDL Foundation concentrates on specifying and evolving the standard, and on maintaining the conformance to the standard through rigorous testing and auditing of procedures. But an ECDL deployment program must also focus on the availability of training courses in the education system, in industry and in independent training schools. It is the availability of opportunities to follow ECDL training courses, which energizes the deployment program. Course vendors are happy with the concept as it gives them a focus for their course offerings. Governmental, employer and trade union organizations in many countries find the concept attractive. Ministries of Education in a number of countries have given active support.

The ECDL Operator usually develops the market by approving its own local test centres throughout the territory. These include private training organizations and schools, in-company training units and state-run training centres, schools, universities, and vocational colleges. Each is chosen for its rigorous adherence to ECDL international testing standards. Nominated staff of a test centre are trained as approved test supervisors of the test centre. The ECDL Operator sells ECSCs to approved companies (in-company programs), training and test centres, for resale or distribution. An ECDL is awarded to an individual when all the modules of the skills card have been certified.

The ECDL Operators control the number of ECDLs issued in their territories. They are also responsible for maintaining the ECDL standard in their country. The contract between the ECDL Foundation and the ECDL Operator grants rights to new versions of the European syllabus and EQTB, and specifies the standards and procedures that have to be adhered to. Conformance to the ECDL standards is controlled through the quality assurance procedures of the ECDL Foundation.

The ECDL Standards and Quality Guidelines cover the following issues:
- Effective administration
- Obligations of the ECDL Foundation to the ECDL Operator
- Obligations of the ECDL Operator to Test Centres
- Test Centre approval procedures
- Testing procedures for a Test Centre

THE DEVELOPMENT OF THE ECDL STANDARD

The initial geographical extent of the ECDL market was defined as the 31 European countries, with a total population of about 500 million. A European target group for the ECDL of roughly 60 million is assumed under current demographic conditions. This is an estimate of the number of individuals who employ IT at a level where it constitutes a necessary element in their working activities. The initial long-term goal for ECDL marketing is to reach about 10 million of the 60 million by year 2005, corresponding to a 17% penetration of today's target group. It is assumed that the public education systems will supply the work force with another 15 - 20 million skilled entrants. The target for 2001 is to have 1.8 million European Computer Skill Cards or Driving licences issued.

The Europe-wide ECDL deployment program, since its launch in 1996/97, has had some successes. Seventeen European countries have initiated ECDL programs and Finland has deployed the CDL. South Africa and Zimbabwe have initiated ICDL programs and now Australia and Canada have announced their participation. The following table, showing figures at the end of March 1999, illustrates the success of the deployment program:

YEAR	SKILLS CARDS DISTRIBUTED
1996	10,000
1997	57,575
1998	180,220
1999 (Target)	330,000

TOP COUNTRIES	SKILLS CARDS
Sweden	177,175
Denmark	61,232
Norway	25,495
Ireland	17,891

At the end of March 1999, 320,173 ECSCs have been distributed, clearly showing that the 1999 target will easily be superceded. Over 100,000 European Computer Driving Licences have been awarded. Sweden has had the most successful deployment program. Sweden has been the first country to introduce automatic testing. The Commission of the European Union has supported ECDL dissemination in the DGIII ESPRIT program, with the DGV European Social Fund and in the DGXXII LEONARDO program. In addition to support from the European Commission and National Government Departments, the ECDL project has created great interest in many large multinational organizations, these include ABB, IBM, Microsoft and Norsk Hydro.

CONCLUSION

The establishment of the ECDL and now the ICDL as a standard of basic competence for personal computer users is well on target. There is wide general acceptance that the ECDL concept is a good idea that meets a need. More than 350,000 people are participating in ECDL programs. There is a large international interest in the ECDL concept. The evolution of the ECDL to the ICDL is being actively supported by the ECDL Foundation. The ICDL is a reality with the launch of deployment programs in South Africa, Zimbabwe, Australia and Canada. With this kind of take up the aspiration of establishing a 'de facto' standard is more than a pipe dream. The take up of ECDL programs is constrained by the training infrastructure and the testing infrastructure. To attempt to establish the ECDL concept as the global standard of IT competency for the individual citizen more inclusive delivery mechanisms are needed to imaginatively distribute training to the needy and deserving current nonuser. To support certification, a readily available and sensitive method of automatic testing and dispensing test results, leading to a full driving licence, is also needed. This points to Internet support for training and automatic testing. This is currently being developed.

GLOSSARY OF TERMS:
(Some elementary definitions are introduce here to support the Chapter.)

Knowledge: Fact of knowing a thing, a state, a person or a truth.

Skill: Practical knowledge in combination with ability to perform Tasks using the basic functionality of a personal computer and its software. A Skill is mastered when the Task is performed at a certain level of competence.

Competence: Implies the attainment of the required standard. Competence in a module of the ECDL is the attainment of a prescribed standard of knowledge or skill in that module.

Task: A piece of work to be done using the basic functionality of a personal computer and its software. The deliverable or outcome of carrying out a Task is an identifiable output from the computer system, an identifiable change of state of the processed document, message, data structure or in the computer or its interface. Consequently, the quality and correctness of the deliverable of the task and the fluency (in terms of time taken) with which the Task is undertaken can be used to measure the performance of a Task (Manual testing or task oriented 'in application' testing). A Task Item (an atomic task) is the atomic keyboard action of a Task (usually a single keyboard action but sometimes a simple keyboard sequence of a simple task).

ANNEX 1. ECDL Syllabus (Part of the Draft Syllabus Version 3.0).
The ECDL Syllabus can be viewed @ (www.ecdl.com)

Module 3 – Word Processing

The following is the Syllabus for Module 3, which is the basis for the Practical Test on Word Processing. The module consists of two sections; the first covering the basics tasks of word processing the second covering some of the more advanced tasks. The candidate must demonstrate the practical ability to use word processing on a computer. The test consists of several tasks covering the basic and advanced tasks.

Goals

The individual shall have a sound knowledge of how a word processing application can be used. He or she shall be able to understand and accomplish normal everyday operations – editing or creating new documents and be able to use the functions available.

Module 3

Category	Skill Set	Ref.	Task Items
3.1 Getting Started	3.1.1 First Steps with Word Processing	3.1.1.1	Open a word processing application.
		3.1.1.2	Open an existing document - make some modifications and save.
		3.1.1.3	Open several documents.
		3.1.1.4	Create a new document.
		3.1.1.5	Use Undo and Repeat commands.
		3.1.1.6	Save a document onto the hard disk or onto a diskette.
		3.1.1.7	Close the document.
		3.1.1.8	Use Online Help functions.
	3.1.2 Change Basic Settings	3.1.2.1	Change display modes.
		3.1.2.2	Use the page view magnification tool/zoom tool.
		3.1.2.3	Modify toolbar display.
3.2 Basic Operations	3.2.1 Insert Data	3.2.1.1	Insert character, word, sentence, or small amount of text.
		3.2.1.2	Insert special characters/symbols.
		3.2.1.3	Insert a page break into a document.
	3.2.2 Select Data	3.2.2.1	Select character, word, sentence, paragraph or entire document.

			Extend or reduce a selection.
	3.2.3 Copy, Move, Delete	3.2.3.1	Use Copy & Paste tools to duplicate text within a document. Use Cut and Paste tools to move text within a document.
		3.2.3.2	Copy and move text between active documents.
		3.2.3.3	Delete text.
	3.2.4 Search & Replace	3.2.4.1	Use the Search command for a word or phrase or piece of text.
		3.2.4.2	Use the Replace command for a word or phrase or piece of text.
3.3 Formatting	3.3.1 Text Formatting	3.3.1.1	Change fonts: sizes and types.
		3.3.1.2	Use italics, bold, underlining, case changes, subscript, superscript.
		3.3.1.3	Use the highlight button.
		3.3.1.4	Apply different colours to text.
		3.3.1.5	Add borders to a document.
		3.3.1.6	Use alignment and justification options.
		3.3.1.7	Indent text.
		3.3.1.8	Change line spacing.
		3.3.1.9	Use lists (bulleted and numbered).
		3.3.1.10	Copy the formatting from a selected piece of text.
		3.3.1.11	Adjust basic tab settings.
3.4 Finishing a Document	3.4.1 Styles and Pagination	3.4.1.1	Apply existing styles to a document.
		3.4.1.2	Insert page numbering in a document.
	3.4.2 Headers & Footers	3.4.2.1	Add Headers and Footers to a document.
		3.4.2.2	Insert date, reference code, author,

		page numbers etc. in Headers and Footers.
	3.4.2.3	Format the contents of Headers and Footers.
3.4.3 Vocabulary & Grammar	3.4.3.1	Use a spell-check program and make changes where necessary.
	3.4.3.2	Specify language.
	3.4.3.3	Use the Grammar tool.
3.4.4 Document Setup	3.4.4.1	Modify document setup, (page orientation, page type, etc.)
	3.4.4.2	Change document margins.
3.5 Printing 3.5.1 Prepare to Print	3.5.1.1	Preview a document.
	3.5.1.2	Print a document from an installed printer.
3.6 Advanced 3.6.1 Tables Functions	3.6.1.1	Create standard tables.
	3.6.1.2	Change cells – format, size, colour & other properties.
	3.6.1.3	Insert and delete columns & rows.
	3.6.1.4	Add borders to a table.
	3.6.1.5	Use the automatic table-formatting tool.
3.6.2 Pictures & Images	3.6.2.1	Add autoshapes to a document: change line colours, change back ground colours.
	3.6.2.3	Move drawn objects or images within a document.
	3.6.2.4	Resize a graphic.
3.6.3 Importing Objects	3.6.3.1	Import a spreadsheet into a document.
	3.6.3.2	Import an image file or graphs into a document.

	3.6.3.3	Import other application files into a document e.g. a presentation document.
3.6.4 Mail Merge	3.6.4.1	Create a mailing list or other data file for use in a Mail Merge.
	3.6.4.2	Merge a mailing list with a letter document or a label document.
3.6.5 Templates	3.6.5.1	Choose an appropriate document template.
	3.6.5.2	Work within a template on a specific task.
3.6.6 Document Exchange	3.6.6.1	Save an existing document under another format: ascii, rtf, version number, etc.
	3.6.6.2	Save a document in a format appropriate for transmission via e-mail.
3.6.7 Hyphenation	3.6.7.1	Use hyphenation in a document.
3.6.8 Tab Settings	3.6.8.1	Use tab commands (centre, decimal, left, right).
3.6.9 Integrated Software	3.6.9.1	Use software which integrates word processing with spreadsheet, database or graphics software

REFERENCES

Dolan, D. (1997). The European Computer Driving Licence. In *Capacity Building for Information Technology in Education in Developing Countries*. Proceedings of CapBIT '97. IFIP Working Groups 3.1, 3.4, 3.5 Working Conference. Harare, Zimbabwe. 25 - 29 August, 1997

European Commission (1996). *The Information Society... and the Citizen*. Brussels: Commission of the European Communities Services. Also http://www.ispo.cec.be

Green Paper (1996). *Living and Working in the Information Society: People First*. COM (96) 389 final. Brussels: Commission of the European Communities Services.

Sherwood-Smith, M.H. (1998) An International Computer Driving Licence. In: Banerjee, P., Hackney, R., Dhillon, G. and Jain, R. (eds). *Business Information Technology Management: Closing the International Divide*, (pp 216-29). New Delhi: Har-Anand,

WRC (1998) Getting Connected: Social Inclusion and the Information Society. Dublin: WRC Social and Economic Consultants Ltd.

ECDL (1997) http://www.ecdl.com

About the Authors

Kai Jakobs has been head of technical staff of the Computer Science Department, Chair of Informatik IV, of the Technical University of Aachen (RWTH) since 1987, and was a member of the university's technical staff from 1985 to 1987. He is the (co-)author of a text book on data communications and of 100+ technical papers on various topics, including naming and addressing, directory and messaging services and, more recently, on different aspects of standards and standardisation processes. He holds a PhD in Computer Science from the University of Edinburgh.

***

Jaroslav Blaha, born in the Czech Republic, is Senior Information Systems Engineer in the Project Implementation Division of the NATO ACCS Management Agency in Brussels. He worked as chief instructor at the German Air Force Technical Academy, and as project manager for communications and information systems at the NATO Headquarters in the Netherlands. Before joining NACMA he was Senior Analyst for mission-critical real-time systems at the NATO Programming Centre in Belgium. He holds university degrees in computer science and economics. He is a permanent member of the NATO COE working group since its establishment, and NACMA's lead for open system engineering. Special interests are software engineering, standardisation and human-machine interaction.

Peter Buxmann is Assistant Professor at the Institute of Information Systems, Frankfurt University, Germany. His current research interests include the impact of web-based emerging technologies like XML on organizations, Supply Chain Management Software, Economics of Standards, and the evaluation of Information Systems, especially business software solutions. Peter Buxmann holds a doctoral degree from the Department of Economics at Frankfurt University. Peter Buxmann can be reached at pbuxmann@wiwi.uni-frankfurt.de. Information about his work, teaching, and project activities is available at http://www.wiwi.uni-frankfurt.de/~pbuxmann.

Tineke Mirjam Egyedi was employed from 1990 to 1994 by the Delft University of Technology as a Ph.D. researcher on standardisation of telematic services. As a consultant for KPN Research, she compared JTC 1 and Internet standardisation. Subsequently, she was engaged by the Royal Institute of Technology in Stockholm to study the role of containers in the transportation infrastructure from a standardisation perspective. She received her Ph.D. degree in 1996. As an employee of the Delft University of Technology, she wrote two environmental policy documents for the Dutch ministry of Transport and Water Works, and studied the use of multimedia in an interdisciplinary group. She participated in two

European multimedia projects (ELECTRA and SLIM) for the University of Maastricht. She is co-editor of 'Social Learning: Multimedia in Education' (in press). Her main interest is standardisation and the development of infrastructures.

Vladislav V. Fomin is a Ph.D. student at the Faculty of Information Technology, COMAS graduate school in computing and mathematical sciences at the University of Jyväskylä, Finland. Prior to 1997 he was at Unibank of Latvia. Current research interests include standard making process in the field of mobile telecommunications, and the social shaping of technology theory. He received Dipl.Eng.oec and MBA degrees from the University of Latvia, is a member of the International Project Management Association (Russian branch).

Wilhelm Hasselbring's research interests include software engineering for cooperative information systems in various application domains. Since 1998 he is Assistant Professor at the INFOLAB, Department of Information Management, Tilburg University. Formerly, he was researcher in Software Engineering at the Universities of Essen and Dortmund. He received his Diploma in Computer Science from the Technical University of Braunschweig in 1989; Ph.D. in Computer Science from the University of Dortmund in 1994. In between, he visited Trinity College in Dublin and the University of Edinburgh. He is a member of ACM, ACM SIGMOD, ACM SIGSOFT, IEEE Computer Society, GI, and GMDS.

Eric J. Iversen is a researcher at the STEP-Group in Oslo, Norway. He has worked principally on the 'framework conditions' of technological change within the tradition of Innovation Studies. His particular interests have been with the changing orientation and importance of Intellectual Property regimes and Standardization. His work has included quantitative work on patents as technology indicators as well as qualitative work. He has a AB from the University of Chicago (AB '90) and a MA from the University of Oslo/ Research Center for Information-technologies and Law (CRID, FUNDP) in Namur, Belgium, where he began looking at the IPR-problem.

Wolfgang König is Professor of the Institute of Information Systems at Frankfurt University, Germany. He holds a degree in Business Administration and Business Pedagogics as well as a Ph.D. degree from the Frankfurt University, where he also completed his habilitation. He spent more than two years at the IBM Research Laboratories in San Jose and Yorktown Heights, as well as at the Kellogg Graduate School of Management, Northwestern University, Evanston and at the University of California at Berkeley. Since 1997, Wolfgang König is head of the Interdisciplinary Research Program "Competitive Advantage by Networking" sponsored by the German national science foundation. He serves as Editor-in-Chief of the leading German IS journal "Wirtschaftsinformatik". He is member of the board of external directors of several IS companies. His research interest is in standardization, networking, and group decision support systems. More information at http://www.wiwi.uni-frankfurt.de/~wkoenig.

Simon Lelieveldt is presently senior policy analyst at the Payment Systems Policy Department of De Nederlandsche Bank (DNB) in Amsterdam, the Netherlands. In 1989, he graduated from Twente University as an industrial engineer in Business Administration (specialization Information Technology) and won the Equity & Law Financial Services Award for his thesis. From 1990 to 1995 he worked as a project manager and strategy advisor for the marketing payments department of the Postbank (market share of 50 % in retail payments in the Netherlands). He has joined DNB in October 1995 and is now responsible for policy development concerning pre-paid cards, Internet payment systems and retail payments in general. He was a member of the G-10 task force that produced a report on the security of electronic money (August 1996) and is currently a member of the G-10 Working Group on Retail Payment Systems. Mr Lelieveldt has written a book and several articles on retail banking and electronic payments, covering economic, technical, legal as well as security aspects. He frequently provides technical assistance on electronic money and on retail payments in general.

Kalle Lyytinen is a professor in Information Systems at the University of Jyväskylä, Finland and currently the Dean of the Faculty of Information Technology. He serves on the editorial boards of several leading IS journals including Information Systems Research, Accounting Management and Information Technologies, Requirements Engineering Journal, and Information Systems Journal. He is currently also the Senior Editor of MISQ. He has published over 70 articles and edited or written seven books or special issues. His research interests include information system theories, system design, system failures and risk assessment, computer supported cooperative work, and diffusion of complex technologies.

Petri Mähönen is working as research professor in the Technical Research Center of Finland (VTT) and is also head of networking research department. He has also docentship at the University of Oulu and contract professorship at the University of Genova. He holds degrees of docent, Ph.D. and M.Sc. Before joining VTT, he has been working and studying at the University of Helsinki, Stanford University (CA, U.S.A.), University of Oulu, Academy of Finland, and University of Oxford (Oxford, U.K.). He has also visited several other international universities as research collaborator and has worked extensively as consultant for several multinational companies. He is member of several standardisation groups, fellow of RAS, and member of IEEE, APS, AAS, ISOC and ComSoc. He has published about 100 papers both in journals and international conferences.

Robert Moreton BA(Hons) MTech FBCS is Associate Dean and Professor of Information Systems at the School of Computing and Information Technology, University of Wolverhampton. His research and consultancy activities have focused on the evaluation of methods and management approaches for information systems development. A common theme of the work has been the tracking and evaluation of developments in information technology and forecasting the potential impact of these developments. He has published widely in the field (one book, many refereed articles and conference papers). Prof. Moreton is a Fellow of the British Computer Society. He also serves as: BCS Membership Panel Assessor (1998-present); Member, Editorial Advisory Board, Journal of Applied Management Studies, Carfax (1995-present); Member, SACWG/HEQC Working Party on Benchmarking Academic Standards (1997-98).

Roy Rada has the following educational credentials: B.A. in Psychology from Yale University in 1973, M.D. in General Medicine from Baylor College of Medicine in 1977, and Ph.D. in Computer Science from Univ. Illinois in Urbana in 1980. He has been involved in standardization in various ways since the mid 1980s when as Editor of Index Medicus he worked on standards for biomedical publications both in paper and online. In the late 1980s and into the mid 1990s Rada was active in CEN and was Project Team leader for the Technical Committee on Medical Informatics. Since the early 1990s, he has written a column for the Communications of the ACM on the topic of standards. His main professional interest is the development of organizations that thrive atop information systems and this interest is intimately linked to the issues of standardization.

Michael Sherwood-Smith PhD, FICS is a Lecturer in Computer Science at University

College Dublin (UCD). He is also working as a Consultant with the ECDL Foundation on a European Commission research project. He worked for 20 years in data processing and management with international companies (Nestlé, W.R. Grace & Unilever). Since 1981 has been in the Computer Science Department at UCD involved in teaching and research and was recently Head of Department. He has been a Project Director on European Commission research projects for over the last sixteen years. His doctoral research was in the field of evaluation of information systems. He has published several papers and co-authored a book, as well as lecturing and consulting in this field. He is a past Chairman of the Irish Computer Society (ICS).

Yesha Y. Sivan is Chief Scientist of The Knowledge Infrastructure Laboratory, Ltd., a start-up company in knowledge management, and a lecturer at Tel Aviv University, where he examines the meanings of the knowledge age. A graduate of Tel Aviv University, he earned his doctorate from Harvard University, where he studied the roles of standards and their potential in linking the cultures of business, knowledge, and technology. His professional experience includes the analysis and development of knowledge systems for corporate, education, and military organizations. For more information, see his home page at www.tau.ac.il/~yesha and his corporate site at www.klab.com.

Michael B. Spring is an Associate Professor of Information Science at the University of Pittsburgh. His research involves the application of technology to the workplace with particular attention large scale electronic document processing and visualization, intelligent agents, and interface design. He received his Bachelor's in Psychology from the College of the Holy Cross, Worcester, MA, and his Ph.D. from the School of Education, University of Pittsburgh. Spring has authored numerous articles and book chapters in the areas of office automation, text and document processing, information technology standardization, and educational technology. He is the author of two book Electronic Printing and Publishing: The Document Processing Revolution and Hands on PostScript.

Klaus Turowski, born 1966, is assistant professor at the University of Magdeburg in the area of Business Information Systems. Before assuming his current position, he worked at the University of Münster. In 1993 he received his Dipl.-Wi.-Ing. (diploma degree in Industrial Engineering and Management) at the University of Karlsruhe and in 1997 his Dr. rer. pol. (Ph.D. in Business Information Systems) at the University of Münster. Since 1998 he is speaker of working group 5.4.6 "Component-oriented Business Application Systems" of German Informatics Society (GI). His research interests are component-oriented software development, production planning and control systems, and Softcomputing techniques.

Martin B.H. Weiss is an associate professor of telecommunications and co-director of the Telecommunications Program at the University of Pittsburgh. He holds a PhD in engineering and public policy from Carnegie Mellon University (1988), where he studied the standards development process. He also holds an MSE in computer, control and information engineering from the University of Michigan (1979) and a BSE in electrical engineering from Northeastern University (1978). His principal research activities have focussed on the issues surrounding the development and adoption of technical compatibility standards.

Tim Weitzel is a researcher and doctoral student at the Institute of Information Systems, Frankfurt University, Germany. His current research interests include economics of standards in information networks and especially the impact of coordination designs on the selection of communication standards. His research activities include an analysis of the standardization problem as part of the Interdisciplinary Research Program "Competitive Advantage by Networking" funded by the German National Science Foundation as well as projects in this area with several companies. He is head of the XML Competence Center (http://xml.cnec.org/). He holds a degree from the Department of Business Administration at Frankfurt University. More information at http://www.wiwi.uni-frankfurt.de/~tweitzel.

Joel West is a researcher at the Center for Research on Information Technology and Organizations at the University of California, Irvine. His research centers on strategies for managing innovation and technological change in global I.T. industries, with a particular focus on the U.S. and Japan. Prior to that, he spent more than a decade as a programmer and manager for commercial software products. He is also a former computer magazine columnist and the author of a book on computer programming.

Falk von Westarp is a researcher and doctoral student at the Institute of Information Systems, Frankfurt University, Germany. His current research interests include coordination in networks, management of software standards in enterprises, and diffusion of innovations in the software market. In the context of these research areas, he conducted various empirical studies (surveys and case studies) and was involved in several projects with companies like Microsoft, Siemens, and Lufthansa. His teaching activities include seminars about software standardization in enterprises and programming courses on Electronic Commerce. Falk von Westarp holds a degree from the Department of Business Administration at Frankfurt University. More information at http://www.wiwi.uni-frankfurt.de/~westarp.

Index

www.ingramcontent.com/pod-product-compliance
Lightning Source LLC
Chambersburg PA
CBHW082005190326
41458CB00010B/3076